Molecular Simulation in Interface and Surfactant

Molecular Simulation in Interface and Surfactant

Editors

Shiling Yuan
Heng Zhang

MDPI • Basel • Beijing • Wuhan • Barcelona • Belgrade • Manchester • Tokyo • Cluj • Tianjin

Editors
Shiling Yuan
Schoole of Chemistry and
Chemical Engineering
Shandong University
Jinan
China

Heng Zhang
Schoole of Chemistry and
Chemical Engineering
Shandong University
Jinan
China

Editorial Office
MDPI
St. Alban-Anlage 66
4052 Basel, Switzerland

This is a reprint of articles from the Special Issue published online in the open access journal *Molecules* (ISSN 1420-3049) (available at: www.mdpi.com/journal/molecules/special_issues/surfactant_simulation).

For citation purposes, cite each article independently as indicated on the article page online and as indicated below:

LastName, A.A.; LastName, B.B.; LastName, C.C. Article Title. *Journal Name* **Year**, *Volume Number*, Page Range.

ISBN 978-3-0365-7471-4 (Hbk)
ISBN 978-3-0365-7470-7 (PDF)

© 2023 by the authors. Articles in this book are Open Access and distributed under the Creative Commons Attribution (CC BY) license, which allows users to download, copy and build upon published articles, as long as the author and publisher are properly credited, which ensures maximum dissemination and a wider impact of our publications.

The book as a whole is distributed by MDPI under the terms and conditions of the Creative Commons license CC BY-NC-ND.

Contents

Heng Zhang, Jiyong Zheng, Cunguo Lin and Shiling Yuan
Molecular Dynamics Study on Properties of Hydration Layers above Polymer Antifouling Membranes
Reprinted from: *Molecules* **2022**, *27*, 3074, doi:10.3390/molecules27103074 1

Shasha Liu, Shiling Yuan and Heng Zhang
Molecular Dynamics Simulation for the Demulsification of O/W Emulsion under Pulsed Electric Field
Reprinted from: *Molecules* **2022**, *27*, 2559, doi:10.3390/molecules27082559 17

Ding Li, Shuixiang Xie, Xiangliang Li, Yinghua Zhang, Heng Zhang and Shiling Yuan
Determination of Minimum Miscibility Pressure of CO_2–Oil System: A Molecular Dynamics Study
Reprinted from: *Molecules* **2021**, *26*, 4983, doi:10.3390/molecules26164983 33

Yingbiao Xu, Yefei Wang, Tingyi Wang, Lingyu Zhang, Mingming Xu and Han Jia
Demulsification of Heavy Oil-in-Water Emulsion by a Novel Janus Graphene Oxide Nanosheet: Experiments and Molecular Dynamic Simulations
Reprinted from: *Molecules* **2022**, *27*, 2191, doi:10.3390/molecules27072191 47

Xiaoheng Geng, Changjun Li, Lin Zhang, Haiying Guo, Changqing Shan and Xinlei Jia et al.
Screening and Demulsification Mechanism of Fluorinated Demulsifier Based on Molecular Dynamics Simulation
Reprinted from: *Molecules* **2022**, *27*, 1799, doi:10.3390/molecules27061799 61

Menghua Li, Haixia Zhang, Zongxu Wu, Zhenxing Zhu and Xinlei Jia
DPD Simulation on the Transformation and Stability of O/W and W/O Microemulsions
Reprinted from: *Molecules* **2022**, *27*, 1361, doi:10.3390/molecules27041361 81

Sergey A. Khrapak
Self-Diffusion in Simple Liquids as a Random Walk Process
Reprinted from: *Molecules* **2021**, *26*, 7499, doi:10.3390/molecules26247499 93

Mohamad Nizam Othman, Alias Jedi and Nor Ashikin Abu Bakar
MHD Stagnation Point on Nanofluid Flow and Heat Transfer of Carbon Nanotube over a Shrinking Surface with Heat Sink Effect
Reprinted from: *Molecules* **2021**, *26*, 7441, doi:10.3390/molecules26247441 101

Tingyi Wang, Hui Yan, Li Lv, Yingbiao Xu, Lingyu Zhang and Han Jia
Computational Investigations of a pH-Induced Structural Transition in a CTAB Solution with Toluic Acid
Reprinted from: *Molecules* **2021**, *26*, 6978, doi:10.3390/molecules26226978 113

Fengfeng Gao
Adsorption of Mussel Protein on Polymer Antifouling Membranes: A Molecular Dynamics Study
Reprinted from: *Molecules* **2021**, *26*, 5660, doi:10.3390/molecules26185660 125

Article

Molecular Dynamics Study on Properties of Hydration Layers above Polymer Antifouling Membranes

Heng Zhang [1], Jiyong Zheng [2], Cunguo Lin [2] and Shiling Yuan [1,*]

1. Key Laboratory of Colloid and Interface Chemistry, Shandong University, Jinan 250199, China; zhangheng@sdu.edu.cn
2. State Key Laboratory for Marine Corrosion and Protection, China Shipbuilding Industry 725 Research Institute, Qingdao 266071, China; zhengjy@sunrui.net (J.Z.); lincg@sunrui.net (C.L.)
* Correspondence: shilingyuan@sdu.edu.cn

Abstract: Zwitterionic polymers as crucial antifouling materials exhibit excellent antifouling performance due to their strong hydration ability. The structure–property relationship at the molecular level still remains to be elucidated. In this work, the surface hydration ability of three antifouling polymer membranes grafting on polysiloxane membranes Poly(sulfobetaine methacrylate) (T4-SB), poly(3-(methacryloyloxy)propane-1-sulfonate) (T4-SP), and poly(2-(dimethylamino)ethyl methacrylate) (T4-DM) was investigated. An orderly packed, and tightly bound surface hydration layer above T4-SP and T4-SB antifouling membranes was found by means of analyzing the dipole orientation distribution, diffusion coefficient, and average residence time. To further understand the surface hydration ability of three antifouling membranes, the surface structure, density profile, roughness, and area percentage of hydrophilic surface combining electrostatic potential, RDFs, SDFs, and noncovalent interactions of three polymers' monomers were studied. It was concluded that the broadest distribution of electrostatic potential on the surface and the nature of anionic SO_3^- groups led to the following antifouling order of T4-SB > T4-SP > T4-DM. We hope that this work will gain some insight for the rational design and optimization of ecofriendly antifouling materials.

Keywords: antifouling polymer; zwitterionic; surface hydration; molecular dynamics simulation

1. Introduction

The adsorption and accumulation of fouling organisms on surface of materials, i.e., marine biofouling, is a major problem faced by ships and offshore facilities [1,2]. The annual cost of increased fuel consumption, cleaning, maintenance, and repair of ships caused by marine biofouling is as high as billions of dollars [3,4]. Early marine antifouling coatings mainly used biotoxic tributyltin (TBT) antifouling paints, which killed marine organism larvae or spores through the release of antifouling agents to achieve antifouling purposes [5,6]. However, traditional antifouling paints are highly toxic for many aquatic organisms and have caused severe damage to the environment. The development of ecofriendly antifouling coatings is gradually becoming a research hotspot in this field [7–9].

Among them, protein-resistant antifouling material that inhibits the settlement of proteins is a relatively promising one [10], such as poly (ethylene glycol) (PEG), zwitterionic polymers [11] (poly (Sulfobetaine methacrylate), pSBMA, or poly (Carboxybetaine methacrylate), pCBMA. For example, Jiang's group [12–14] has been engaged in biofouling research for a long period and synthesized a series of zwitterionic polymers. On the one hand, they used molecular simulation methods to reveal the antifouling mechanism of materials on the microscopic level. On the other hand, they carried out application research on this basis to design and synthesize new antifouling materials. Zheng and coworkers [15–17] investigated the antifouling properties of zwitterionic polymer brushes, polyacrylamide, and hydroxyalkyl acrylamides using combined molecular dynamics and steered molecular dynamics, believing that the carbon space and anionic groups have distinct effects on their antifouling performance. The state key laboratory of marine corrosion and protection in China has also synthesized a series of antifouling coatings by grafting zwitterionic sulfobetaine methacrylate (T4-SB) or anionic sulfonate methacrylate (T4-SP), which have the property of inhibiting adsorption of proteins on the surface of polysiloxane material (T4). These materials have a good antifouling effect on fouling organisms such as diatoms. We found that the static adsorption number of diatoms in the T4-SP antifouling material is $15/mm^2$ (4% of the T4 antifouling material) in the experiment; for T4-SB, the static adsorption number of diatoms is $9/mm^2$ (2% of the T4 antifouling material), which significantly improved the antifouling performance of the silicone material.

The adsorption of protein on surface is affected by many factors [18–21], among which the factors favorable for adsorption mainly include the enthalpy loss from the van der Waals and electrostatic attraction between protein and surface, and the entropic gain from the removal of hydration layer at the surface of material and protein. The disadvantages include the enthalpy gain required for the dehydration of surface and protein, protein's conformation adjustment, as well as the entropic loss from protein adsorption and exposure of hydrophobic regions. The hydration layer above the surface of the antifouling material plays a crucial role from the antifouling perspective [22] because it provides the physical and energy barriers that must be overcome during protein adsorption. To confirm the structure of the hydration layer above the surface of antifouling materials, many experimental studies have been carried out. For example, Leng et al. [23,24] confirmed that there is a tightly bounded and regularly ordered hydration layer above zwitterionic antifouling membrane compared with polymer membrane without antifouling ability using sum frequency generation (SFG) vibrational spectroscopy. Paul et al. [25] directly observed the structure of hydration layer above the surface of epoxy organosilane modified silica nanoparticles and unmodified silica nanoparticles by frequency modulation—atomic force microscopy. Combined with molecular dynamics simulations, a more continuous and thicker hydration layer structure was found on the surface of modified silica particles, which endows the material with a better antifouling ability.

In this work, we will compare the antifouling ability of three polymer antifouling membranes (T4-DM, T4-SP, T4-SB) using molecular dynamics simulation at the molecular level through the hydration layer. We hope this work will provide theoretical support for the subsequent design and optimization of related antifouling materials.

2. Simulation Method

2.1. Model

Three antifouling membranes were constructed according to their molecular structures (Figure 1). The T4 substrate was neglected considering the main differences between different antifouling membranes focusing on the grafted polymers. The modeling process of T4-DM system is illustrated in Figure 2 as an example. The polymer chains with a degree of polymerization of 15 (Figure 2b) were built from their repeat unit (Figure 2a) using the Visualizer module in Materials Studio. This was repeated 10 times in the x and y directions to derive the initial configuration of antifouling membrane in Figure 2c. The initial configurations of the antifouling membranes were then subject to a 21-step molecular dynamics compression and relaxation [26] to obtain the equilibrium packing structure (which might not be the optimal one) in Figure 2d. The procedure of the 21-step MD simulation protocol is listed in Table S1. The simulation boxes were then enlarged two times along the z-axis to accommodate solvent molecules (Figure 2e). As a comparison, antifouling membranes without water were also studied (Figure 2g). Finally, all systems were subject to equilibrium molecular dynamics simulations to derive equilibrium structures (Figure 2f,h).

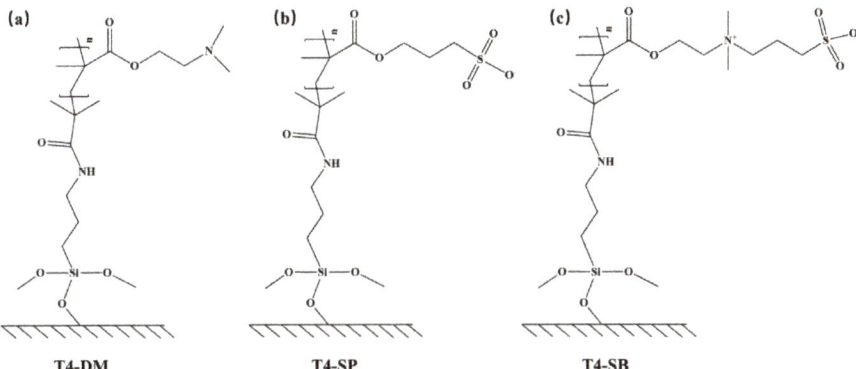

Figure 1. Chemical structure of three nonfouling membranes (**a**) T4-DM, (**b**) T4-SP, (**c**) T4-SB.

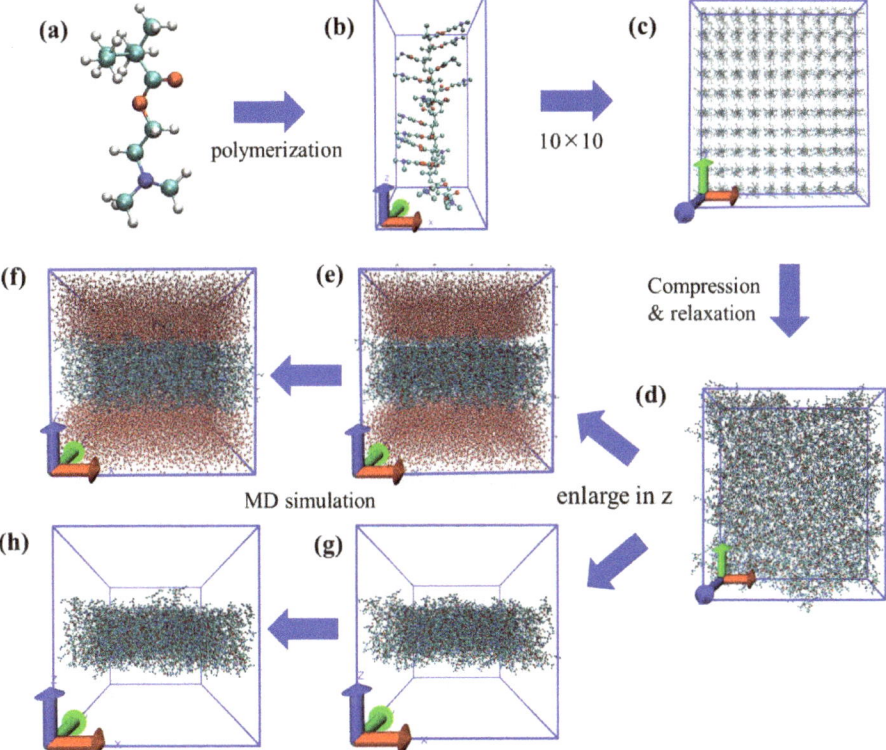

Figure 2. Modeling process and Simulation protocol of T4-DM system. (**a**) Repeat unit of DM; (**b**) single polymer chain of DM in simulation box (side view), (**c**) enlarged 10 times in x and y directions of (**b**) (top view); (**d**) compressed and relaxed configuration of DM membrane (top view); (**e**) initial configuration of DM with water system (side view); (**f**) final configuration of DM with water system (side view); (**g**) initial configuration of DM without water system (side view); (**h**) final configuration of DM without water system (side view).

2.2. Simulation Details

The repeat unit of each polymer was calculated at B3LYP/def2SVP// B3LYP/def2TZVP level using Gaussian 16 [27]. Then, RESP charges were derived from Multiwfn 3.8 [28]. All molecular dynamics simulations were performed using Gromacs 2019.3 software package [29]. Gromos 54a7 force field was used [30]. The total potential energy was given as a combination of valence terms, including bond stretching, angle bending, torsion, and nonbonded interactions. The nonbonded interactions between atoms were described by the Lennard-Jones potential, and the standard geometric mean combination rules were used for the van der Waals interactions between different atom species. Water molecules used the SPC model [31].

In the simulations, each of the systems was initialized by minimizing the energies of the initial configurations using steepest descent method. Following the minimization, a 50 ns MD simulation under NPT ensemble was carried out for each system, with a time step of 2 fs. In all simulations, the temperature was kept constant at 298 K by the v-rescale thermostat algorithm [32]. The pressure was kept constant at 1 atm by the Berendsen algorithm [33]. Bond lengths were constrained using the LINCS algorithm and periodic boundary conditions were applied in all directions [34]. Short-range nonbonded interactions were cut off at 1.2 nm, with long-range electrostatics calculated using the particle mesh Ewald method [35]. Trajectories were stored every 2 ps and visualized using VMD 1.9.3 [36].

3. Results and Discussion

3.1. Properties of Antifouling Membranes

3.1.1. Density Profiles

The simulated configurations of three antifouling membranes at dry and hydrated states are illustrated in Figure S1. We can clearly see that there are no significant differences between T4-DM membrane under dry and hydrated states, while for T4-SP and T4-SB membranes, many side chains extend to water phase. This indicates that the side chains of T4-SP and T4-SB have a better hydrophilicity. Besides this, the compression of these chains during adsorption of foulant would reduce the conformation possibility, which is entropically unfavorable, subsequently causing steric repulsion and preventing adsorption [10].

To quantitatively study the structure of three antifouling membranes, the density profile along z-axis was calculated, as shown in Figure 3. The results were derived from the last 5 ns trajectory. The density profile was symmetrized around the membrane center to obtain a better result. The density profiles of T4-DM in dry and hydrated states almost overlapped. As for T4-SP and T4-SB membranes, the density profile of hydrated state broadened compared with that of dry state (more obvious for T4-SB membrane), which is consistent with the configurations in Figure S1. The density of water in T4-SB is higher than that of T4-SP, and even higher than that of T4-DM, which indicates that the side chains of T4-SB can attract extra water molecules compared with those of T4-SP and T4-DM. We then deduce that the hydrophilicity of the antifouling membranes follows the order of T4-SB > T4-SP > T4-DM.

3.1.2. Surface Roughness

Since the density profile is a statistical average of the entire membrane layer, it cannot reflect the local specific structural information of membranes. To further analyze the detailed surface structure, contour maps of the upper surface of three antifouling membranes in hydrated states were sketched, as shown in Figure 4. To define the surface of membrane, the simulation box was divided into grids with 0.4 nm × 0.4 nm resolution in xy plane. Atoms with the largest or smallest z-axis were selected as the top atoms to define the membrane surface. It can be seen from Figure 5 that T4-DM membrane's surface is relatively flat, while T4-SP has more peaks and valleys than T4-DM. As for T4-SB, the contour lines are the densest, indicating that the order of surface roughness is T4-SB > T4-SP > T4-DM.

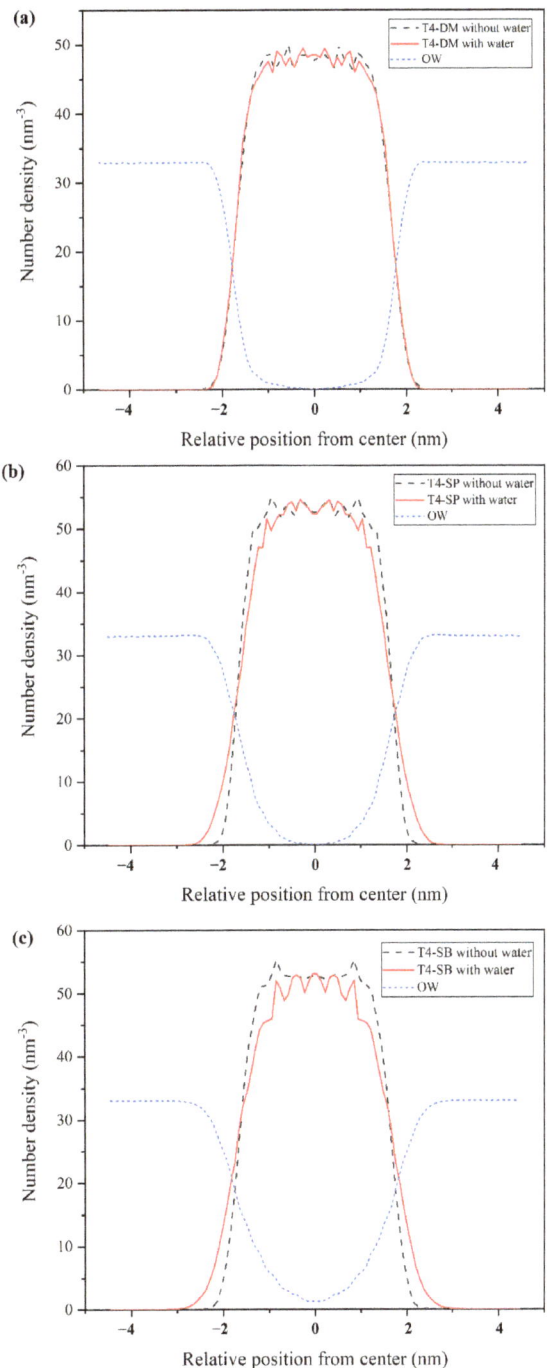

Figure 3. Density profiles along z-axis. (**a**) T4-DM, (**b**) T4-SP, (**c**) T4-SB. Black dashed lines and red lines represent density of antifouling membranes under dry state or hydrated states. Dotted blue lines represent density of water.

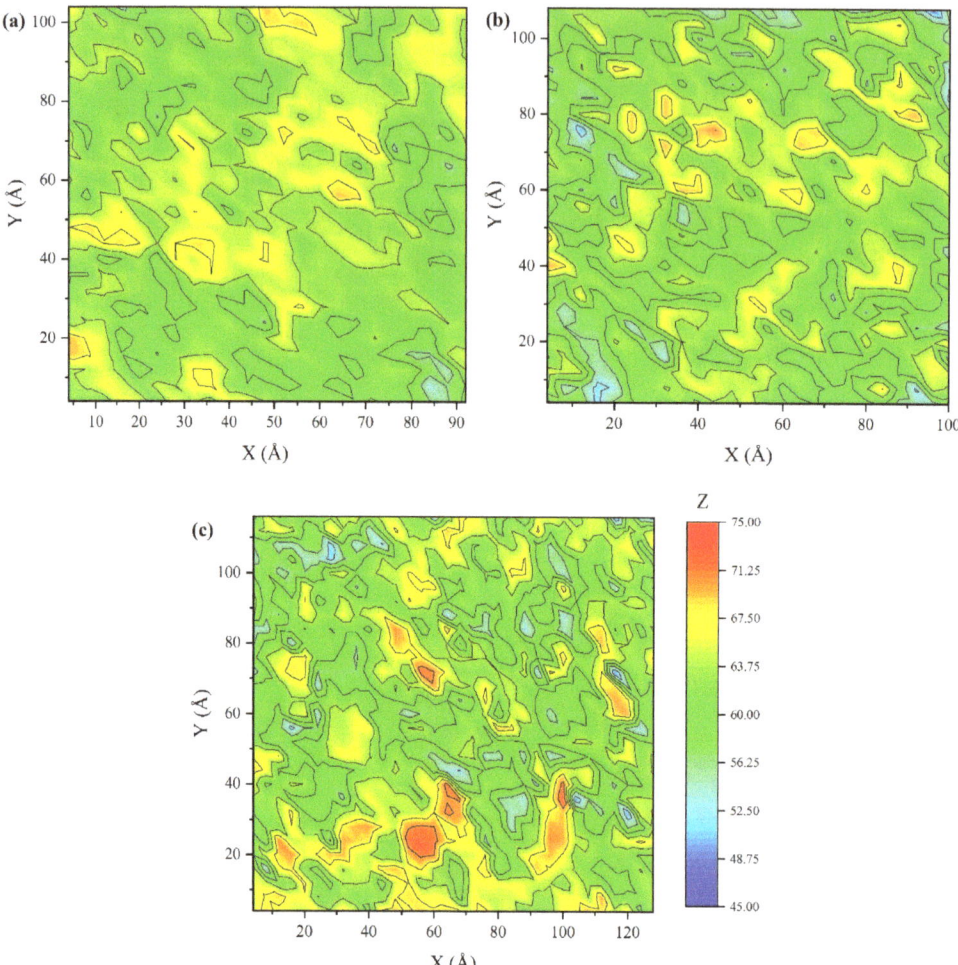

Figure 4. Contour maps of three antifouling membrane surfaces. (**a**) T4-DM, (**b**) T4-SP, (**c**) T4-SB.

To quantify the surface roughness of three antifouling membranes, the root mean square roughness R was introduced [37]:

$$R = \sqrt{\frac{\sum_{i=1}^{N}(Z_i - \overline{Z})^2}{N}}$$

where Z_i is the z-coordinate of the atoms exposed in the outermost layer in each grid point and $\overline{Z}$ is the average value of the z-coordinates of all the atoms exposed on the outermost surface. Both the up and down surfaces of three antifouling membranes in dry and hydrated states are calculated and listed in Table 1. The data suggested there is little difference between dry and hydrated states for T4-DM. The roughness in hydrated state follows the order of T4-SB > T4-SP > T4-DM, which is consistent with Figures 3 and 5. Obviously, the greater the roughness of the surface, the more hydrophilic sites were exposed, and the more water molecules could be bound.

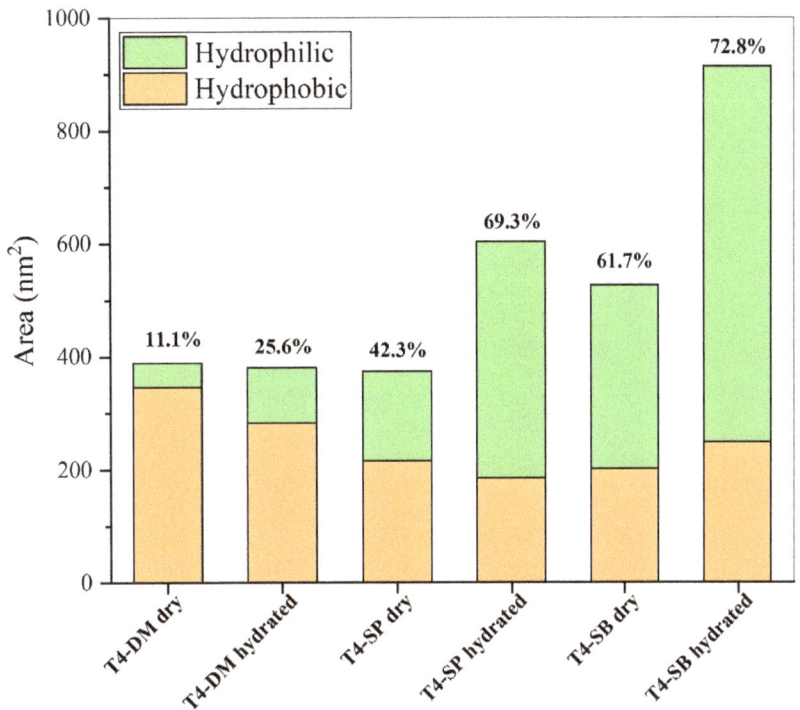

Figure 5. Solvent accessible surface area including hydrophobic and hydrophilic part of three antifouling membranes. Numbers above the bar means the proportion of hydrophilic area.

Table 1. Root-mean-square roughness of three antifouling membranes.

	Root-Mean-Square Roughness R *			
	Hydrated		Dry	
	Up	Down	Up	Down
T4-DM	2.80 ± 0.074	2.84 ± 0.062	2.79 ± 0.056	2.80 ± 0.068
T4-SP	3.73 ± 0.050	3.83 ± 0.056	2.68 ± 0.037	2.41 ± 0.037
T4-SB	4.28 ± 0.068	4.20 ± 0.059	2.94 ± 0.042	2.98 ± 0.032

* Data derived from the last 1 ns trajectory.

3.1.3. Hydrophilicity

In addition to the influence of surface roughness on surface hydration, the hydrophilicity and hydrophobicity of the surface determine the surface hydration ability directly. The hydrophilic and hydrophobic surface area of each antifouling membrane were calculated from the last 5 ns trajectory, as shown in Figure 5. During calculation, the atomic charge between −0.2 and 0.2 was considered as the hydrophobic surface area, and the other is the hydrophilic surface area. The hydrophilic surface area and its proportion of all three antifouling membranes increased in hydrated state. The total surface area does not change much between dry and hydrated states, which is consistent with the configuration in Figure S1. The total surface area, especially the hydrophilic surface area, of T4-SP and T4-SB both increased significantly when immersed in water, which suggests that they have a strong hydration ability.

3.2. Properties of Surface Hydration Layer

3.2.1. Structural Properties

After the above structural analysis of the antifouling membranes, it was found that the surface hydration ability of the three antifouling membranes was T4-SB > T4-SP > T4-DM. We also noticed that with the increase in surface hydration ability, more water molecules can penetrate into the matrix of membrane from the density profiles in Figure 3. To examine the structure of water molecules near the interface of antifouling membranes, we calculated the cosine of the angle between dipole of water and z-axis at different distances from the surface, as shown in Figure 6. Obviously, for a random distribution, the cosθ should be close to 0 [38]. In the T4-DM membrane system, only water molecules close to membrane have a certain orientation, while water molecules farther away are randomly distributed. In the T4-SP system, the dipole orientation of surface water molecules slightly decreased to 0 after 2 nm, while in the T4-SB system, there is still a long-distance arrangement of water molecules even beyond 2 nm away from the surface. This observation is consistent with Leng's experiment [23,24], where ordered water molecules were found at zwitterionic pSBMA surfaces.

3.2.2. Dynamic Properties

The antifouling membranes can also affect the hydration layer's dynamic properties beside the structure of surface water molecules. We calculated the distribution of the average residence time of water molecules within 0.5 nm of antifouling membrane surfaces, as shown in Figure 7b. The average residence time means how long water molecules can stay near the surface of the antifouling membrane on average [39]. It reflects the stability of the hydration water layer of the antifouling membrane or, in other words, the hydration ability of antifouling membranes [40]. Figure 7a shows the trajectory of one hydration layer water molecule above T4-SB membrane. The calculated average residence time is shown in Table 2. It can be seen that the average residence time increased from T4-DM and T4-SP to T4-SB, indicating that the binding effect of antifouling membranes on their surface hydration layers increased.

The diffusion behavior of surface hydration layer water molecules above three antifouling membranes was investigated. The mean square displacement (MSD) of surface hydration layer water molecules is shown in Figure 8. Their diffusion coefficients were then calculated according to Einstein's equation and collected in Table 2. It can be seen that the diffusion coefficients of surface hydration layer water molecules above three antifouling membranes gradually decreased from T4-DM and T4-SP to T4-SB, indicating that the mobility of water molecules decreased or the binding effect from the antifouling membranes increased, which is consistent with the previous analysis.

Table 2. Dynamic properties of hydration layer water molecules above three antifouling membranes including average residence time and diffusion coefficient.

Antifouling Membranes	Average Residence Time (ps)	Diffusion Coefficient D $\times$ 10^{-5} cm^2/s
T4-DM	17.85	2.57 (+/− 0.080)
T4-SP	24.98	1.62 (+/− 0.014)
T4-SB	27.16	1.54 (+/− 0.043)
Bulk water	-	4.13 (+/− 0.15)

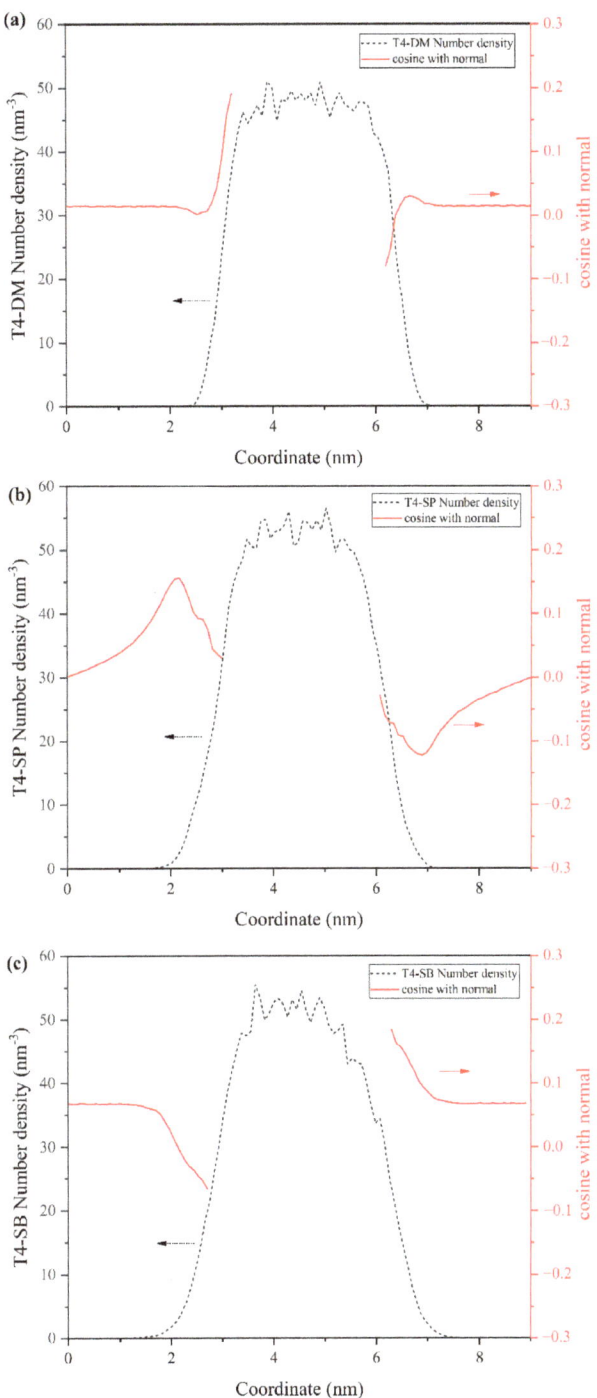

Figure 6. Water dipole orientation profiles of three antifouling membranes. (**a**) T4-DM, (**b**) T4-SP, (**c**) T4-SB.

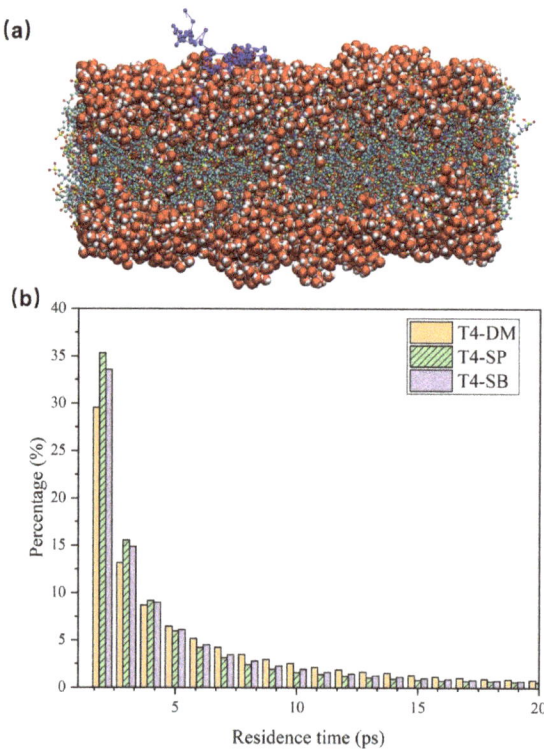

Figure 7. Trajectory and residence time of surface hydration layer water molecules. (**a**) Trajectory of one hydration layer water molecule above T4-SB surface (connected blue dots). The antifouling membrane was colored in CPK mode. The surface hydration layer water molecules were modeled in VDW mode. (**b**) Residence time distribution of water molecules in the hydration layer of three antifouling membranes.

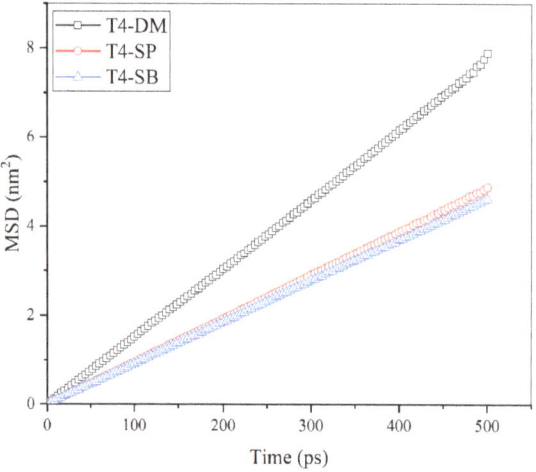

Figure 8. Mean square displacement of water molecules in the hydration layer of nonfouling membrane.

3.3. Hydration Mechanisms—From the View of Monomers

3.3.1. Solvation Free Energy

We have analyzed and compared the structural properties of the antifouling membranes and the structural and dynamic properties of their hydration water layers from the overall antifouling membranes' view. The order of surface hydration ability or antifouling ability, T4-SB > T4-SP > T4-DM, was obtained. Next, we analyze the mechanisms for the difference in hydration ability from the monomer's view, which serves as a model for the antifouling polymer membrane [38].

Solvation free energies were calculated for three monomers at M05-2X/6-31 g* level, as collected in Table 3. The negative of solvation free energy indicates all three monomers have a high affinity with water. The order of solvation free energy follows the order of T4-SB > T4-SP >> T4-DM, which is consistent with previous analysis.

Table 3. Properties of monomer of three antifouling membranes.

Monomer of Antifouling Membranes	Solvation Free Energy (kcal/mol)	Nonpolar Surface Area ($Å^2$)	Polar Surface Area ($Å^2$)	Molecular Polarity Index (kcal/mol)	Number of Bonded Water Molecules
T4-DM	−6.16	203.59	63.16	8.58	10.02
T4-SP	−71.68	0.00	283.56	67.07	15.43
T4-SB	−73.46	24.60	339.35	43.54	18.43

3.3.2. Electrostatic Potential

Electrostatic potentials of the three monomers were calculated and mapped on their van der Waals surfaces [41], as shown in Figure 9. The molecular polarity, polar, and nonpolar surface area were also calculated, as shown in Table 3 [42]. The surface area with |ESP| <= 10 kcal/mol was considered as nonpolar surface area while the others were considered as polar surface area. It can be seen that the negative charge center of T4-DM monomer is located at the N atom. Since T4-SP monomer has a negative charge, the overall electrostatic surface is negative, and mainly concentrated on the sulfonate group. In the zwitterion T4-SB monomer, the negative charge center is located in the sulfonate group and the positive charge center is located at the N atom. Though the MPI of T4 SP was the largest, the T4-SB has the largest polar surface area, which can combine with more water. Combining with the distribution of areas occupied by different electrostatic potential regions in Figure 9b, it can be seen that the distribution of electrostatic potential on the surface of T4-SB monomer is the broadest, which is conducive to the electrostatic interaction with other polar molecules such as water [43].

3.3.3. Radial Distribution Function

To further understand the hydration ability of antifouling polymers' monomers, another molecular dynamics simulation was conducted. Three monomers were solvated in $4 \times 4 \times 4$ nm^3 water box, respectively; then, 50 ns NPT simulations were performed. After that, the radial distribution functions (RDFs) of the water molecules or Na^+ around the polar groups of three monomers and their coordination number were calculated, respectively, as shown in Figure 10. The RDFs can reflect the intermolecular structure and interactions between center atoms and surrounding water molecules. Two peaks were found in the RDF curve, indicating that two hydration layers were formed, which corresponded to the first hydration layer that consisted of bound water and the second hydration layer made up of trapped water; this agrees with Paul's experiment [25]. According to Figure 10a,b, SO_3^- groups in T4-SP and T4-SB have similar hydration ability and are stronger than the N group in T4-DM and T4-SB. Meanwhile, the peaks of $g(r)_{N-OW}$ in T4-DM were lower than those in T4-SB and also the coordination number of the first hydration shell from Figure 11c,d, indicating a better packed hydration shell around N in T4-SB. The number of water molecules tightly bonded to three monomers were also calculated and collected in Table 3. Consequently, the T4-SB antifouling membrane presents a more hydrophilic behavior than T4-SP and T4-DM.

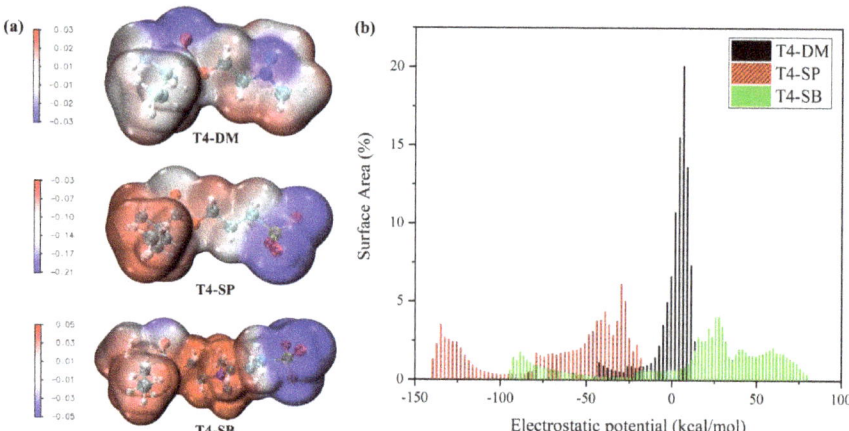

Figure 9. Electrostatic potential of monomers of three antifouling polymers' monomers. (**a**) Electrostatic potential mapped on vdW surface; (**b**) distribution of surface area percentage of different electrostatic potentials.

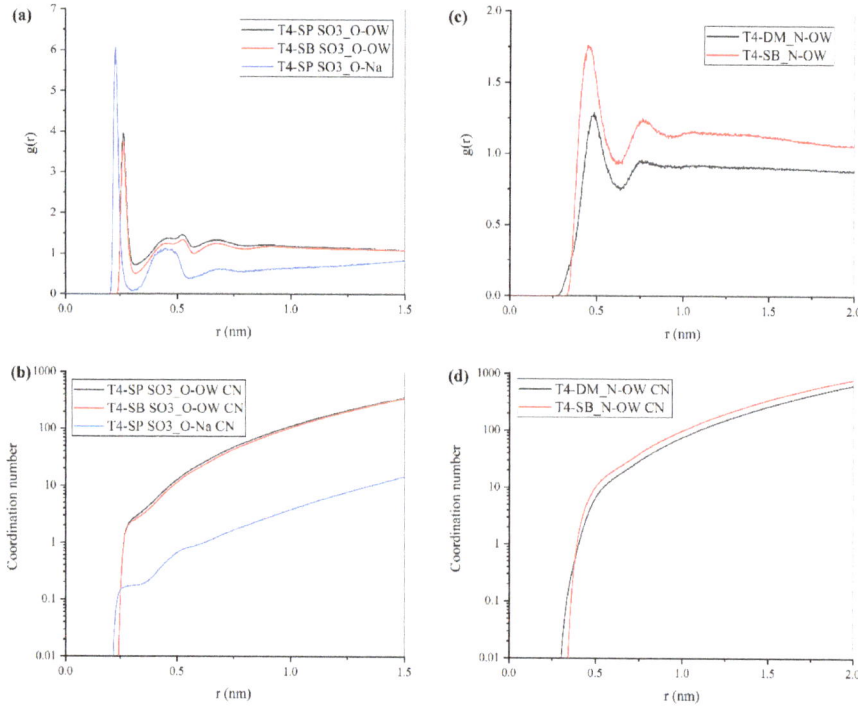

Figure 10. Radial distribution functions (RDFs), g(r) of hydration groups on antifouling membrane surface, and the cumulative number. (**a**) RDF of SO_3^-_O-OW and SO_3^-_O-Na^+, oxygens in SO_3^- groups as referenced atoms. (**b**) RDF of N-OW, nitrogen atoms as referenced atoms. (**c**) Cumulative number of SO_3^-_O-OW and SO_3^-_O-Na^+. (**d**) Cumulative number of N-OW.

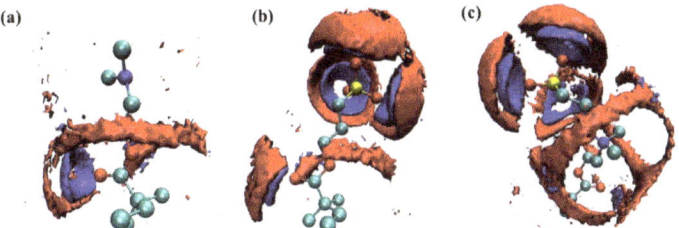

Figure 11. Spatial distribution function of water molecules around three different antifouling monomers. (**a**) DM, (**b**) SP, (**c**) SB.

3.3.4. Spatial Distribution Function

Though the RDFs can reflect the hydration effect of hydrophilic groups in three monomers on water molecules, the calculation of RDFs is based on the spherical averaging of the water molecules around the hydrophilic group, which neglects the spatial distribution of the water molecules. Therefore, the spatial distribution function (SDF) of water molecules around hydrophilic groups was calculated, shown in Figure 11. From this, we can see that there is only a ribbonlike distribution around the carbonyl oxygen in DM monomer, while the distribution of water molecules around the N atom cannot be shown under current isosurface. In the SP monomer, there are three spherical crown water molecule distribution areas in the direction of three S–O bonds, which is obviously caused by the hydrogen bond formed between the O atom in SO_3^- group and the water molecules. Similar structures were also found in SB monomer. Besides this, there is a ribbonlike distribution of water molecules around the N atom.

3.3.5. Noncovalent Interactions

To fundamentally understand the different hydration ability of three antifouling monomers, aNCI (averaged noncovalent interaction) analysis [44,45] was conducted, shown in Figure 12. The green area in the figure indicates that van der Waals interaction is dominant. Blue area indicates that there is a strong hydrogen bond interaction. The red area indicates that there is a strong steric hindrance effect. In DM monomer, as the negative charge center N atom was shielded by surrounding methyl groups, it can only interact with water molecules through weak vdW interactions. In T4-SP and T4-SB monomers, water molecules can directly form hydrogen bonds with the exposed O atoms, which plays a key role in their strong hydration ability. Besides that, the extra positive charge center N atom can also interact with water molecules through weak vdW interactions such as N in the T4-DM monomer. Therefore, the hydration abilities of three antifouling polymers are in the order of T4-SB > T4-SP > T4-DM.

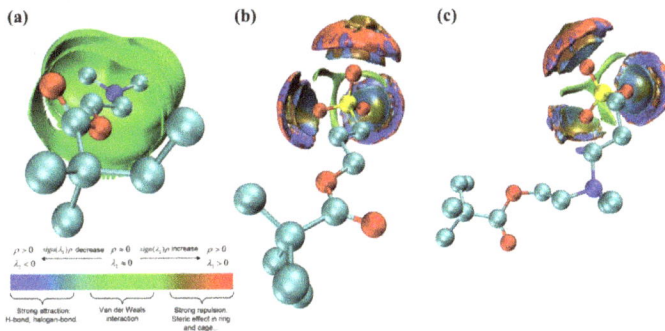

Figure 12. Noncovalent interaction around three different nonfouling repeat units. (**a**) DM, (**b**) SP, (**c**) SB.

4. Conclusions

In this work, we investigated the surface hydration of three antifouling membranes—T4-DM, T4-SP, and T4-SB—by a series of molecular dynamics simulations. Dipole orientation distribution, diffusion coefficient, and average residence time revealed an orderly, packed, and tightly bound surface hydration layer above T4-SP and T4-SB antifouling membranes. The surface structure, density profile, surface roughness, and area percentage of hydrophilic surface provide further details regarding the strong surface hydration of T4-SP and T4-SB from the membranes' aspect. The side chains of T4-SP and T4-SB were more stretched in hydrated state due to their high hydration ability, which can cause steric repulsion and prevent adsorption. Their surfaces are relatively rough, which can bind much more water or even let water penetrate into the internal voids of the membrane.

To further understand the surface hydration ability of three antifouling membranes, solvation free energy, electrostatic potential, RDFs, SDFs, and noncovalent interactions of three monomers were analyzed. T4-SB monomer has the broadest distribution of electrostatic potential on the surface, resulting from the separated negatively and positively charge center and largest water coordination number for its zwitterionic architecture. Its exposed negative charge center SO_3^- group can form hydrogen bonds with surrounding water molecules and the shielded positive charge center N can also bind water molecules through weak vdW interaction.

The simulation data suggest the hydration ability of monomers ranks in terms of T4-SB > T4-SP > T4-DM. Since the surface hydration layer serves as a physical and energy barrier during the prevention of protein adsorption, we believe their antifouling ability ranks in terms of T4-SB > T4-SP > T4-DM, which is consistent with experiments.

Supplementary Materials: The following supporting information can be downloaded at: https://www.mdpi.com/article/10.3390/molecules27103074/s1, Figure S1: Final simulation configuration of three antifouling membranes under dry and hydrated states; Table S1: 21-step MD compression and relaxation schemes.

Author Contributions: Conceptualization, H.Z., J.Z. and S.Y.; methodology, H.Z.; software, H.Z.; validation, H.Z.; formal analysis, H.Z.; investigation, H.Z.; writing—original draft preparation, H.Z.; writing—review and editing, S.Y.; supervision, S.Y.; project administration, S.Y.; funding acquisition, S.Y. and C.L. All authors have read and agreed to the published version of the manuscript.

Funding: This research was funded by the open fund of State Key Laboratory of Marine Corrosion and Protection grant number 6142901190402, and Natural Science Foundation of Shandong Province grant number ZR2021MB055.

Institutional Review Board Statement: Not applicable.

Informed Consent Statement: Not applicable.

Data Availability Statement: Not applicable.

Acknowledgments: We gratefully appreciate the financial support from the open fund of State Key Laboratory of Marine Corrosion and Protection (6142901190402) and Natural Science Foundation of Shandong Province (ZR2021MB055).

Conflicts of Interest: The authors declare no conflict of interest.

Sample Availability: Samples of the compounds are not available from the authors.

References

1. Pradhan, S.; Kumar, S.; Mohanty, S.; Nayak, S.K. Environmentally Benign Fouling-Resistant Marine Coatings: A Review. *Polym. Technol. Mater.* **2018**, *58*, 498–518. [CrossRef]
2. Hu, P.; Xie, Q.; Ma, C.; Zhang, G. Silicone-Based Fouling-Release Coatings for Marine Antifouling. *Langmuir* **2020**, *36*, 2170–2183. [CrossRef] [PubMed]
3. Callow, M.E.; Callow, J.E. Marine biofouling: A sticky problem. *Biologist* **2002**, *49*, 10–14. [PubMed]
4. Yeon, D.K.; Ko, S.; Jeong, S.; Hong, S.-P.; Kang, S.M.; Cho, W.K. Oxidation-Mediated, Zwitterionic Polydopamine Coatings for Marine Antifouling Applications. *Langmuir* **2019**, *35*, 1227–1234. [CrossRef]

5. Uc-Peraza, R.G.; Castro, Í.B.; Fillmann, G. An absurd scenario in 2021: Banned TBT-based antifouling products still available on the market. *Sci. Total Environ.* **2022**, *805*, 150377. [CrossRef]
6. Ali, A.; Jamil, M.I.; Jiang, J.; Shoaib, M.; Amin, B.U.; Luo, S.; Zhan, X.; Chen, F.; Zhang, Q. An overview of controlled-biocide-release coating based on polymer resin for marine antifouling applications. *J. Polym. Res.* **2020**, *27*, 85. [CrossRef]
7. Nurioglu, A.G.; Esteves, A.C.C.; de With, G. Non-toxic, non-biocide-release antifouling coatings based on molecular structure design for marine applications. *J. Mater. Chem. B* **2015**, *3*, 6547–6570. [CrossRef]
8. Lejars, M.; Margaillan, A.; Bressy, C. Fouling Release Coatings: A Nontoxic Alternative to Biocidal Antifouling Coatings. *Chem. Rev.* **2012**, *112*, 4347–4390. [CrossRef]
9. Silva, E.R.; Ferreira, O.; Ramalho, P.A.; Azevedo, N.F.; Bayón, R.; Igartua, A.; Bordado, J.C.; Calhorda, M.J. Eco-friendly non-biocide-release coatings for marine biofouling prevention. *Sci. Total Environ.* **2019**, *650*, 2499–2511. [CrossRef]
10. Maan, A.M.C.; Hofman, A.H.; de Vos, W.M.; Kamperman, M. Recent Developments and Practical Feasibility of Polymer-Based Antifouling Coatings. *Adv. Funct. Mater.* **2020**, *30*, 2000936. [CrossRef]
11. Zhang, Y.; Liu, Y.; Ren, B.; Zhang, D.; Xie, S.; Chang, Y.; Yang, J.; Wu, J.; Xu, L.; Zheng, J. Fundamentals and applications of zwitterionic antifouling polymers. *J. Phys. D Appl. Phys.* **2019**, *52*, 403001. [CrossRef]
12. Shao, Q.; Jiang, S. Molecular Understanding and Design of Zwitterionic Materials. *Adv. Mater.* **2015**, *27*, 15–26. [CrossRef] [PubMed]
13. Jiang, S.; Cao, Z. Ultralow-Fouling, Functionalizable, and Hydrolyzable Zwitterionic Materials and Their Derivatives for Biological Applications. *Adv. Mater.* **2010**, *22*, 920–932. [CrossRef] [PubMed]
14. Lin, X.; Jain, P.; Wu, K.; Hong, D.; Hung, H.-C.; O'Kelly, M.B.; Li, B.; Zhang, P.; Yuan, Z.; Jiang, S. Ultralow Fouling and Functionalizable Surface Chemistry Based on Zwitterionic Carboxybetaine Random Copolymers. *Langmuir* **2019**, *35*, 1544–1551.
15. Liu, Y.; Zhang, D.; Ren, B.; Gong, X.; Xu, L.; Feng, Z.-Q.; Chang, Y.; He, Y.; Zheng, J. Molecular simulations and understanding of antifouling zwitterionic polymer brushes. *J. Mater. Chem. B* **2020**, *8*, 3814–3828. [CrossRef]
16. Liu, Y.; Zhang, D.; Ren, B.; Gong, X.; Liu, A.; Chang, Y.; He, Y.; Zheng, J. Computational Investigation of Antifouling Property of Polyacrylamide Brushes. *Langmuir* **2020**, *36*, 2757–2766. [CrossRef]
17. Liu, Y.; Zhang, Y.; Ren, B.; Sun, Y.; He, Y.; Cheng, F.; Xu, J.; Zheng, J. Molecular Dynamics Simulation of the Effect of Carbon Space Lengths on the Antifouling Properties of Hydroxyalkyl Acrylamides. *Langmuir* **2019**, *35*, 3576–3584. [CrossRef]
18. Zhang, P.; Ratner, B.D.; Hoffman, A.S.; Jiang, S. 1.4.3 A-Nonfouling Surfaces. In *Biomaterials Science*, 4th ed.; Wagner, W.R.; Sakiyama-Elbert, S.E., Zhang, G., Yaszemski, M.J., Eds.; Academic Press: Cambridge, MA, USA, 2020; pp. 507–513.
19. Quan, X.; Liu, J.; Zhou, J. Multiscale modeling and simulations of protein adsorption: Progresses and perspectives. *Curr. Opin. Colloid Interface Sci.* **2019**, *41*, 74–85. [CrossRef]
20. Liu, S.; Tang, J.; Ji, F.; Lin, W.; Chen, S. Recent Advances in Zwitterionic Hydrogels: Preparation, Property, and Biomedical Application. *Gels* **2022**, *8*, 46. [CrossRef]
21. Wang, Z.; Scheres, L.; Xia, H.; Zuilhof, H. Developments and Challenges in Self-Healing Antifouling Materials. *Adv. Funct. Mater.* **2020**, *30*, 1908098. [CrossRef]
22. Chen, S.; Li, L.; Zhao, C.; Zheng, J. Surface hydration: Principles and applications toward low-fouling/nonfouling biomaterials. *Polymer* **2010**, *51*, 5283–5293. [CrossRef]
23. Leng, C.; Hung, H.-C.; Sieggreen, O.A.; Li, Y.; Jiang, S.; Chen, Z. Probing the Surface Hydration of Nonfouling Zwitterionic and Poly(ethylene glycol) Materials with Isotopic Dilution Spectroscopy. *J. Phys. Chem. C* **2015**, *119*, 8775–8780. [CrossRef]
24. Leng, C.; Hung, H.-C.; Sun, S.; Wang, D.; Li, Y.; Jiang, S.; Chen, Z. Probing the Surface Hydration of Nonfouling Zwitterionic and PEG Materials in Contact with Proteins. *ACS Appl. Mater. Interfaces* **2015**, *7*, 16881–16888. [CrossRef]
25. Molino, P.J.; Yang, D.; Penna, M.; Miyazawa, K.; Knowles, B.R.; MacLaughlin, S.; Fukuma, T.; Yarovsky, I.; Higgins, M.J. Hydration Layer Structure of Biofouling-Resistant Nanoparticles. *ACS Nano* **2018**, *12*, 11610–11624. [CrossRef] [PubMed]
26. Larsen, G.S.; Lin, P.; Hart, K.E.; Colina, C.M. Molecular Simulations of PIM-1-like Polymers of Intrinsic Microporosity. *Macromolecules* **2011**, *44*, 6944–6951. [CrossRef]
27. Frisch, M.; Trucks, G.; Schlegel, H.; Scuseria, G.; Robb, M.; Cheeseman, J.; Scalmani, G.; Barone, V.; Petersson, G.; Nakatsuji, H.J.W.C. *Gaussian 16 Revision A. 03*; Gaussian Inc.: Wallingford, CT, USA, 2016; Volume 2.
28. Lu, T.; Chen, F. Multiwfn: A multifunctional wavefunction analyzer. *J. Comput. Chem.* **2012**, *33*, 580–592. [CrossRef] [PubMed]
29. Abraham, M.J.; Murtola, T.; Schulz, R.; Páll, S.; Smith, J.C.; Hess, B.; Lindahl, E. GROMACS: High performance molecular simulations through multi-level parallelism from laptops to supercomputers. *SoftwareX* **2015**, *1–2*, 19–25. [CrossRef]
30. Schmid, N.; Eichenberger, A.P.; Choutko, A.; Riniker, S.; Winger, M.; Mark, A.E.; van Gunsteren, W.F. Definition and testing of the GROMOS force-field versions 54A7 and 54B7. *Eur. Biophys. J.* **2011**, *40*, 843. [CrossRef]
31. Berendsen, H.; Postma, J.; Van Gunsteren, W.; Hermans, J. *Intermolecular Forces*; Pullman, B.D., Ed.; Reidel Publishing Company: Dordrecht, The Netherlands, 1981.
32. Bussi, G.; Donadio, D.; Parrinello, M. Canonical sampling through velocity rescaling. *J. Chem. Phys.* **2007**, *126*, 014101. [CrossRef]
33. Berendsen, H.J.C.; Postma, J.P.M.; Van Gunsteren, W.F.; DiNola, A.; Haak, J.R. Molecular dynamics with coupling to an external bath. *J. Chem. Phys.* **1984**, *81*, 3684–3690. [CrossRef]
34. Hess, B.; Bekker, H.; Berendsen, H.J.; Fraaije, J.G. LINCS: A linear constraint solver for molecular simulations. *J. Comput. Chem.* **1997**, *18*, 1463–1472. [CrossRef]

35. Darden, T.; York, D.; Pedersen, L. Particle mesh Ewald: An N·log(N) method for Ewald sums in large systems. *J. Chem. Phys.* **1993**, *98*, 10089–10092. [CrossRef]
36. Humphrey, W.; Dalke, A.; Schulten, K. VMD: Visual molecular dynamics. *J. Mol. Graph.* **1996**, *14*, 33–38. [CrossRef]
37. Jahan Sajib, M.S.; Wei, Y.; Mishra, A.; Zhang, L.; Nomura, K.-I.; Kalia, R.K.; Vashishta, P.; Nakano, A.; Murad, S.; Wei, T. Atomistic Simulations of Biofouling and Molecular Transfer of a Cross-linked Aromatic Polyamide Membrane for Desalination. *Langmuir* **2020**, *36*, 7658–7668. [CrossRef]
38. Huang, H.; Zhang, C.; Crisci, R.; Lu, T.; Hung, H.-C.; Sajib, M.S.J.; Sarker, P.; Ma, J.; Wei, T.; Jiang, S.; et al. Strong Surface Hydration and Salt Resistant Mechanism of a New Nonfouling Zwitterionic Polymer Based on Protein Stabilizer TMAO. *J. Am. Chem. Soc.* **2021**, *143*, 16786–16795. [CrossRef]
39. Zhu, Y.; Lu, X.; Ding, H.; Wang, Y. Hydration and Association of Alkaline Earth Metal Chloride Aqueous Solution under Supercritical Condition. *Mol. Simul.* **2003**, *29*, 767–772. [CrossRef]
40. Předota, M.; Bandura, A.V.; Cummings, P.T.; Kubicki, J.D.; Wesolowski, D.J.; Chialvo, A.A.; Machesky, M.L. Electric Double Layer at the Rutile (110) Surface. 1. Structure of Surfaces and Interfacial Water from Molecular Dynamics by Use of ab Initio Potentials. *J. Phys. Chem. B* **2004**, *108*, 12049–12060. [CrossRef]
41. Zhang, J.; Lu, T. Efficient evaluation of electrostatic potential with computerized optimized code. *Phys. Chem. Chem. Phys.* **2021**, *23*, 20323–20328. [CrossRef]
42. Liu, Z.; Lu, T.; Chen, Q. Intermolecular interaction characteristics of the all-carboatomic ring, cyclo 18 carbon: Focusing on molecular adsorption and stacking. *Carbon* **2021**, *171*, 514–523. [CrossRef]
43. Gu, Q.-A.; Liu, L.; Wang, Y.; Yu, C. Surface modification of polyamide reverse osmosis membranes with small-molecule zwitterions for enhanced fouling resistance: A molecular simulation study. *Phys. Chem. Chem. Phys.* **2021**, *23*, 6623–6631. [CrossRef]
44. Wu, P.; Chaudret, R.; Hu, X.; Yang, W. Noncovalent Interaction Analysis in Fluctuating Environments. *J. Chem. Theory Comput.* **2013**, *9*, 2226–2234. [CrossRef] [PubMed]
45. Johnson, E.R.; Keinan, S.; Mori-Sánchez, P.; Contreras-García, J.; Cohen, A.J.; Yang, W. Revealing Noncovalent Interactions. *J. Am. Chem. Soc.* **2010**, *132*, 6498–6506. [CrossRef] [PubMed]

Article

Molecular Dynamics Simulation for the Demulsification of O/W Emulsion under Pulsed Electric Field

Shasha Liu [1,2], Shiling Yuan [1] and Heng Zhang [1,*]

[1] School of Chemistry and Chemical Engineering, Shandong University, Jinan 250100, China; liushasha325@163.com (S.L.); shilingyuan@sdu.edu.cn (S.Y.)
[2] School of Chemistry and Chemical Engineering, Qilu Normal University, Jinan 250100, China
* Correspondence: zhangheng@sdu.edu.cn

Abstract: A bidirectional pulsed electric field (BPEF) method is considered a simple and novel technique to demulsify O/W emulsions. In this paper, molecular dynamics simulation was used to investigate the transformation and aggregation behavior of oil droplets in O/W emulsion under BPEF. Then, the effect of surfactant (sodium dodecyl sulfate, SDS) on the demulsification of O/W emulsion was investigated. The simulation results showed that the oil droplets transformed and moved along the direction of the electric field. SDS molecules can shorten the aggregation time of oil droplets in O/W emulsion. The electrostatic potential distribution on the surface of the oil droplet, the elongation length of the oil droplets, and the mean square displacement (MSD) of SDS and asphaltene molecules under an electric field were calculated to explain the aggregation of oil droplets under the simulated pulsed electric field. The simulation also showed that the two oil droplets with opposite charges have no obvious effect on the aggregation of the oil droplets. However, van der Waals interactions between oil droplets was the main factor in the aggregation.

Keywords: bidirectional pulsed electric field; O/W emulsion; demulsification; molecular dynamics simulation

1. Introduction

With the increase of oil production activities, oil pollution, particularly oily wastewater, has become an environmental concern nowadays. Enormous quantities of oily wastewater are generated during different industrial processes all around the world, including petroleum refining, industrial discharges, petroleum exploration, food production operations, etc. [1–5]. The oils in wastewater include fats, lubricants, cutting oils, heavy hydrocarbons and light hydrocarbons [6]. These oils can be further divided into free oils and emulsified oils. The free oils in wastewater are easier to separate by physical techniques such as gravity separation and skimming [7,8]. However, the emulsified oil droplets are more difficult to handle due to their high stability in water [9,10]. A widely used separation technique for emulsified oils involves the addition of chemicals, such as ferric or aluminum salts, to induce colloidal destabilization [1,5]. However, the cost is expensive, and the chemicals would dissolve in water or form-settling sludge after the treatment, which is not recommended from the perspective of green chemistry.

An alternative approach is the use of an electric field, especially for the dehydration of crude oil [11–17]. Electric field demulsification has practical advantages such as a lack of extra chemicals, simple equipment, short process flow, etc. It can achieve physical separation of oil and water mixtures and recover oily substances to a certain extent, without the pollution from added chemicals [18,19]. The demulsification mechanism of W/O emulsion by electric fields has also been widely researched. The demulsification was attributed to the droplets' polarization and elongation under the electric field, which then induces interactions between dipoles, leading to aggregation [20–22]. However, the utilization of an electric field to separate oil and water in O/W emulsion is rarely studied. It is generally believed that electric field demulsification does not work for O/W emulsions, because water is conductive, and the electrical energy could dissipate easily in aqueous solution [5]. Ichikawa et al. [23] investigated the demulsification process of dense O/W emulsion in a low-voltage DC electric field and found that a mass of gas bubbles occurred and surged in the emulsion during the demulsification process. Furthermore, Hosseini et al. [24] applied a non-uniform electric field to demulsify the benzene-in-water emulsion. Bubbles were also generated in the emulsion when the electric field was introduced. The occurrence of these phenomena is attributed to the overly large electric current in the emulsion, leading to water's electrolysis.

To resolve this problem, Bails et al. [25,26] applied a pulsed electric field (PEF) to W/O emulsion and found that the electric current generated by a pulsed electric field is small at a high voltage. After this, Ren et al. [27] applied bidirectional pulsed electric field (BPEF) to separate O/W emulsion; this was prepared by mixing 0 # diesel oil and SDS solution. They found that BPEF induced the aggregation of oil droplets, and BPEF had a distinct demulsification effect on O/W emulsion with surfactant. The demulsification effect under different BPEF voltages, frequencies and duty cycles were investigated by evaluating oil content and turbidity of the clear liquid after demulsification. Moreover, they put forward a hypothesis that charges on the oil drop surface would redistribute under BPEF to promote the mutual attraction and coalescence of oil drops. However, the mechanism of oil droplet movement and aggregation in O/W emulsion at the molecular level under BPEF has not been well studied; still less studied is the effect of surfactants on demulsification. Molecular dynamics (MD) simulation is considered a useful tool to carry out microscopic analysis of the dynamic behavior of nanodroplets based on the basic laws of classical mechanics [28]. Chen et al. [29] used MD simulation to study the influence of a direct current electric field on the viscosity of waxy crude oil and the microscopic properties of paraffin. They found that the electric field strength affects the distribution of oil molecules. He et al. [30] simulated the aggregation process and behavior of charged droplets under different pulsed electric field waveforms by MD simulation. They discovered that the deformation of droplets is greatly affected by the waveform. Moreover, the additive in the emulsion has an important influence on its emulsifying stability [31–33]. For example, an experimental study found that BPEF had a distinct demulsification effect on O/W emulsion with SDS surfactant [27] However, to the best of our knowledge, there has been no report on the microscopic level of the demulsification of O/W systems with SDS surfactants under the action of BPEF electric field. In addition, the crude oil composition was relatively distinct when the behavior of crude oil in electric field was simulated previously [34]. Therefore, it is necessary to study the movement and coalescence behavior of oil droplets in O/W emulsion under BPEF by MD simulation. We believe that this will provide a theoretical basis for the application of BPEF in O/W emulsion demulsification.

In this paper, we investigated the movement and aggregation behavior of crude oil droplets in O/W emulsion with differing contents of SDS under BPEF. First, the structural changes of oil droplets in each system and their collision time were analyzed to determine the behavioral difference between the oil droplets with and without SDS. Second, the centroid distance between oil droplets, the average elongation length of oil droplets and the MSD of SDS and asphaltene molecules of oil droplets were calculated to explain why SDS can reduce demulsification time. Finally, we investigated the aggregation behavior of oil droplets after the shut-off of BPEF and discussed the aggregation mechanism of oil droplets under BPEF.

2. Results and Discussion

2.1. Emulsified Crude Oil Droplet

It was believed that SDS increased the hydrophilicity of oil droplets by increasing the hydrophilic surface area of the droplets [34]. In order to study the surface condition of oil droplets with a different SDS content in each system, the solvent-accessible surface area (SASA) was calculated and shown in Figure 1. With an increasing number of SDS molecules, we noted that both the hydrophilic surface and the hydrophobic surface of crude oil droplets also increased. Meanwhile, the ratio of hydrophilic area to hydrophobic area also increased significantly. Therefore, adding SDS molecules could increase the hydrophilic surface area of oil droplets more significantly. In addition, the greater the number of SDS molecules, the greater the hydrophilic surface area.

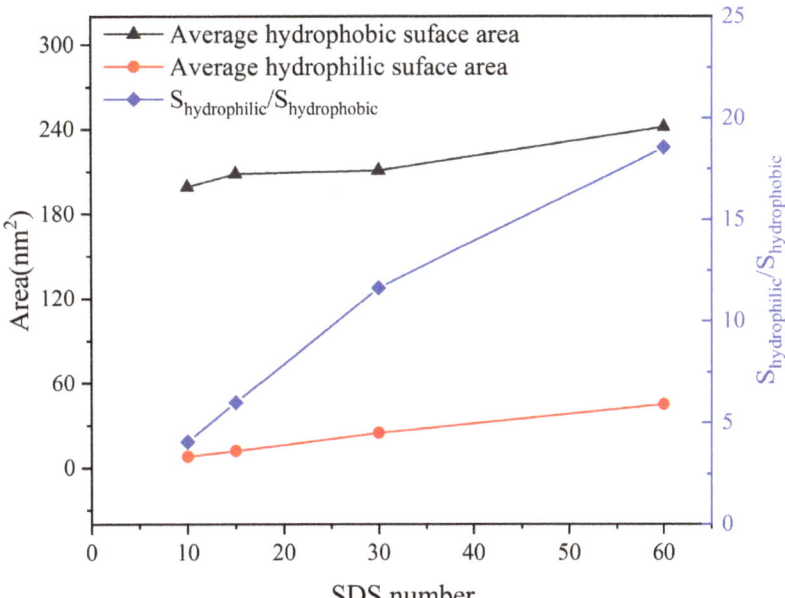

Figure 1. Average hydrophobic and hydrophilic surface areas of emulsified oil droplet.

2.2. Dynamic Behavior of Oil Droplets under BPEF

In order to study the behavior of oil droplets under electric field, BPEF with $E = 0.50$ V/nm was applied in the z-direction of all systems. Figure 2 displayed the conformational changes of oil droplets with differing SDS content during the electric field output stage. As can be seen from Figure 2, all oil droplets gradually deformed under the electric field, elongated in the z-direction and migrated toward the opposite direction of the electric field. Moreover, SDS and asphaltene molecules concentrated at the end of the oil droplet. Excess SDS molecules were distributed along the entire surface of the deformed oil droplets (System IV, V). To clearly see the distribution of SDS and asphaltene, oil droplets of System II were partly zoomed in. It can be seen that the SDS and asphaltene molecules aggregated at the head of the oil droplet's moving direction, with the negative sulfonic acid groups of SDS and carboxyl groups of asphaltene molecules facing the opposite direction of the electric field. Therefore, we thought it was the polar SDS and asphaltene molecules that guided the movement of oil droplets under the electric field.

In Figure 2 we also noted that the states of two oil droplets in five systems were different at 400 ps. In System II and V, the two oil droplets collided at 400 ps. Whereas, for System I, III and IV, this didn't occur. To investigate the impact of SDS concentration on the coalescence of oil droplets driven by electric field, the collision time was summarized in Figure 3. A collision occurred when the minimum distance between two oil droplets was less than 0.35 nm. It was found that the addition of SDS molecules can reduce the collision time of oil droplets, especially with the 6.2% SDS concentration condition

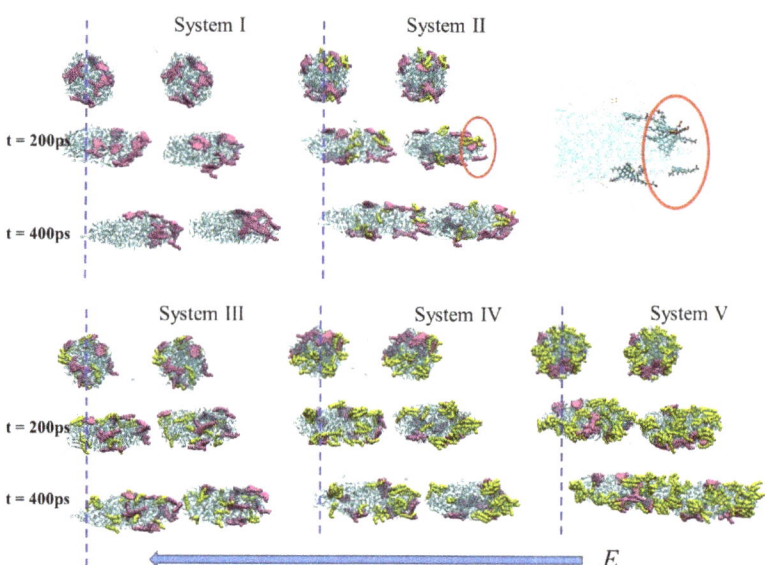

Figure 2. The behavior of oil droplets under BPEF output duration. Asphaltene molecules were colored pink. SDS molecules were colored yellow. Resins and light crude oil molecules were colored green.

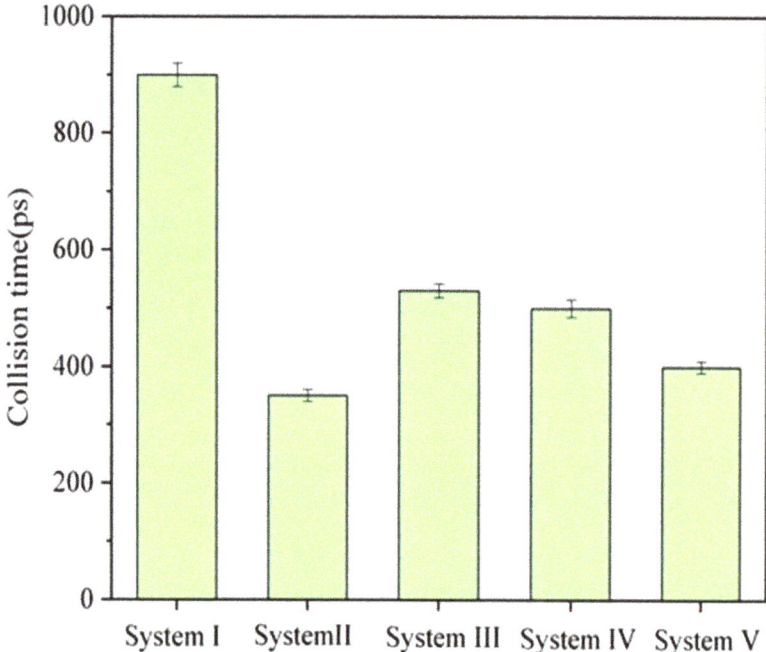

Figure 3. Collision time of oil droplets in five systems under the electric field of 0.5 V/nm.

2.3. Surface Charge Distribution

The electrostatic potential surface of the oil droplet can reflect its charge redistribution under electric field. Considering the two oil droplets in each system are the same, only one droplet's electrostatic potential was calculated. The electrostatic potential diagrams were obtained for different systems at the initial and specific time during the simulation (Figure 4). It can be found that under the influence of the hydrophilic and negatively charged asphaltene molecules on the surface, some areas of the oil droplets appear electronegative (blue area) before electric field is applied. The electronegative area increases with the increase of SDS content. However, the electrostatic potential at the surface of the deformed oil droplet noticeably changed under the electric field, which is manifested as one end of ellipsoidal oil droplet being electronegative toward the opposite direction of the electric field and the other end being electropositive, as in System I and II. These revealed that the redistribution of the charge of oil droplets under electric field resulted in the droplets' polarization. This phenomenon was consistent with the experimental observation that the charge of the oil droplets under the electric field is positive in the direction of the electric field, and negative in the opposite direction. Meanwhile, we found that for Systems III, IV and V, the oil droplets' polarization was not obvious. To explain this, the number density of SDS in oil droplets under electric field was analyzed at the same time. In Figure 5, we defined the middle of the oil drop as 0 and the moving direction as the positive direction. It can be seen that with an increase in SDS, it tended to distribute on the surface of the whole deformed oil droplet, which could further explain why the electronegativity area of the deformed oil droplet increased with the increase in SDS.

The dynamic behavior of SDS and asphaltene molecules of oil droplets and the electrostatic potential distribution on the surface of oil droplets displayed that the mobile negative charges on oil droplets moved toward the opposite direction of the applied electric field. However, what causes the two oil droplets moving in the same direction to collide, and its relationship with SDS content is unclear. Figure 6 presented the centroid distance between two oil droplets and the average elongation length l_e of the two oil droplets along the z direction from the application of the electric field to the collision of oil droplets in each system. We can find that even when the two oil droplets were deformed under electric field, the centroid distance between the two oil droplets remained approximately 10 nm in all systems. This means that due to the oil droplets having the same composition in each system, they moved along the opposite direction of the electric field at almost the same speed, so they kept almost the same initial centroid distance. However, the average elongation length l_e of the two oil droplets in the five systems studied were significantly different. It was found that in all systems, the oil droplets start length was about 6 nm in diameter, and their length increased with time; the average elongation length l_e of the oil droplets exceeded 10 nm near the collision time point. This means that when the length of the oil droplet is stretched enough, the two oil droplets are connected head to end; that is, a collision occurs. Meanwhile, we noted that in Figure 6b, the order of the growth rate of the average elongation length l_{e2} from largest to smallest is System II, System V, System III ≈ System IV and System I, which is similar to the trend showing the variation of the collision time of studied systems in this work. Therefore, for the O/W emulsion system with uniform distribution of oil droplets, we thought the demulsification collision time in the electric field is significantly affected by SDS. Adding appropriate SDS surfactant into O/W systems can effectively reduce the power consumption.

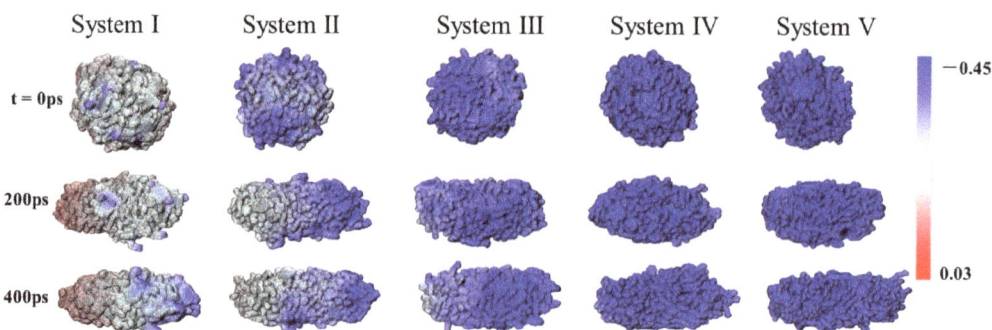

Figure 4. Electrostatic potential surface of the oil droplets during simulation.

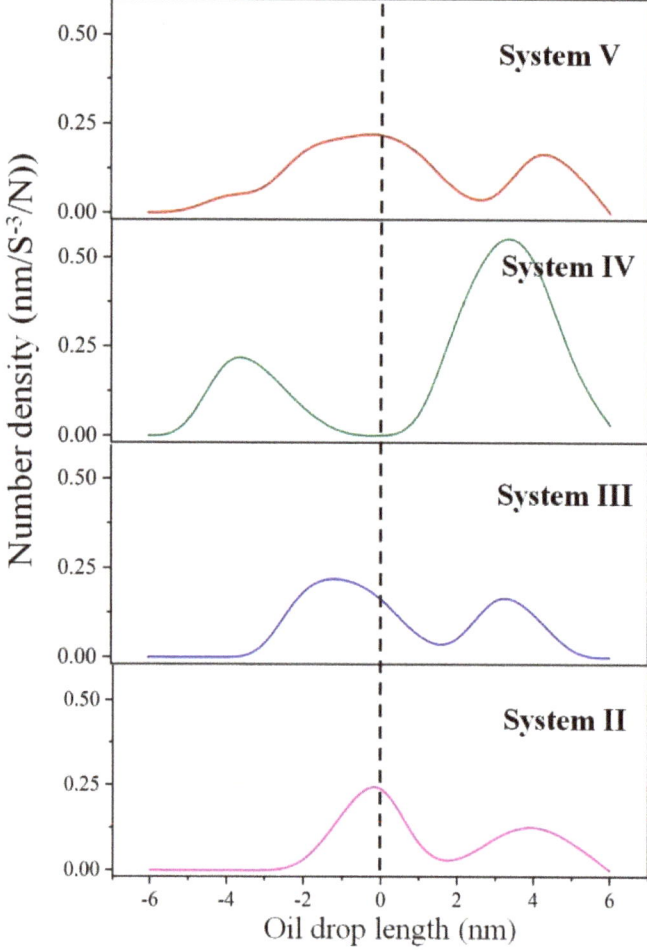

Figure 5. Density profile of SDS in oil droplets under electric field at the same time.

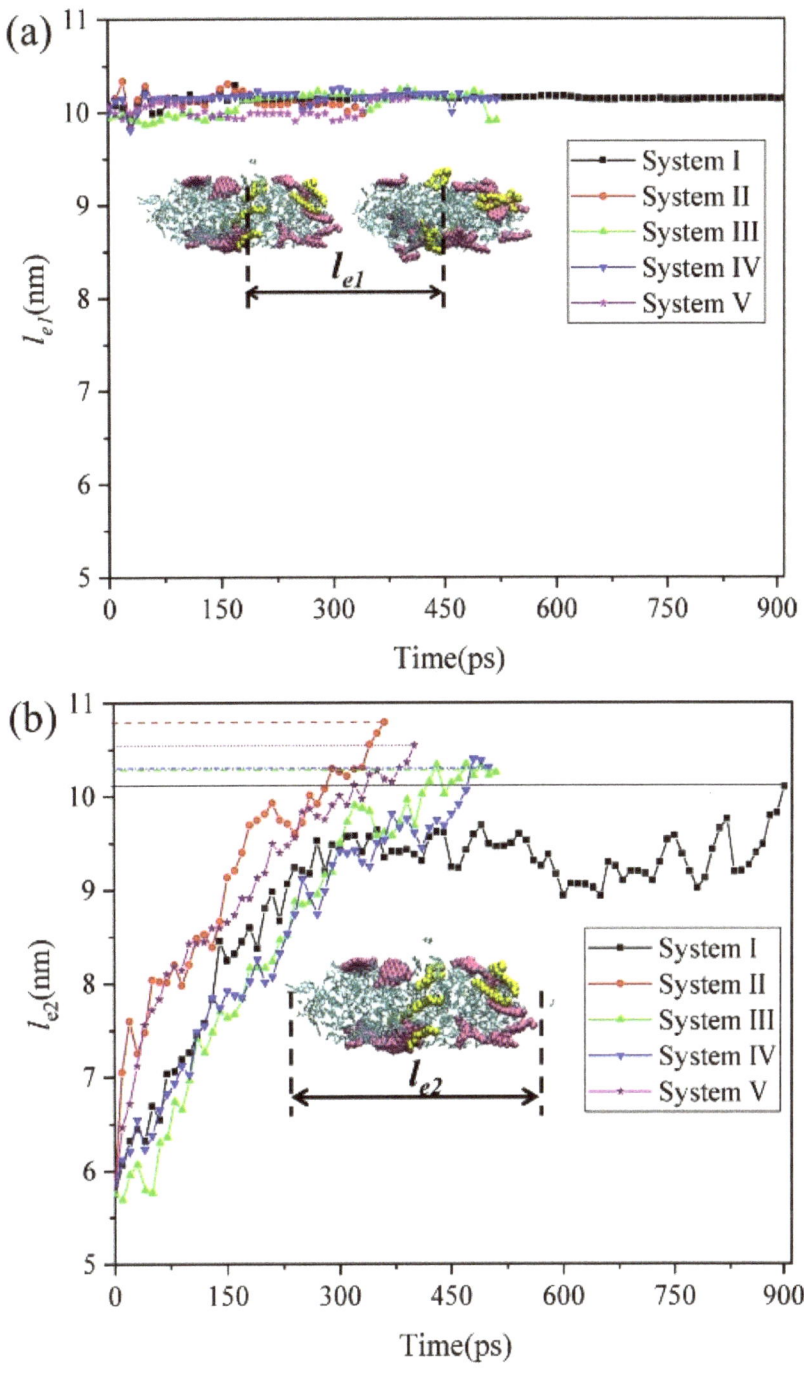

Figure 6. Centroid distance between two oil droplets (**a**) and average elongation length *le* of the two oil droplets along the z direction (**b**).

As discussed above, SDS and asphaltene molecules guide the entire oil droplet to move in the opposite direction of the electric field. It was predicted that the average elongation length of oil droplets in electric field is related to the diffusivity of the SDS and asphaltene molecules. Thus, we calculated the MSD of SDS and asphaltene molecules for the five systems in Figure 7. It was found that the order of SDS and asphaltene molecules' diffusion from largest to smallest was System II, System V, System IV, System III and System IV in the five systems studied; this was consistent with the order of the average elongation length of oil droplets under electric field. We thought that in System I the asphaltene molecules interacted more strongly with the surrounding oil molecules due to the influence of its structure, which decreased its mobility under electric field. However, the negative SDS molecules are smaller and demonstrate strong mobility in the electric field, thus increasing their overall mobility. However, this does not mean that the greater the SDS content in the oil droplets, the greater the mobility of negatively charged molecules. Therefore, the SDS content of the oil droplets have great significance on the demulsification effect. Meanwhile, we calculated the root-mean-square fluctuation (RMSF) of oil droplets during the electric field output stage (as shown in Supplementary Materials: Figure S1). By comparing the RMSF of the three systems, we found that the fluctuation of System II and System IV was stronger than that of System I. The addition of SDS could have accelerated the movement of oil droplets, which was similar to the calculation result of MSD.

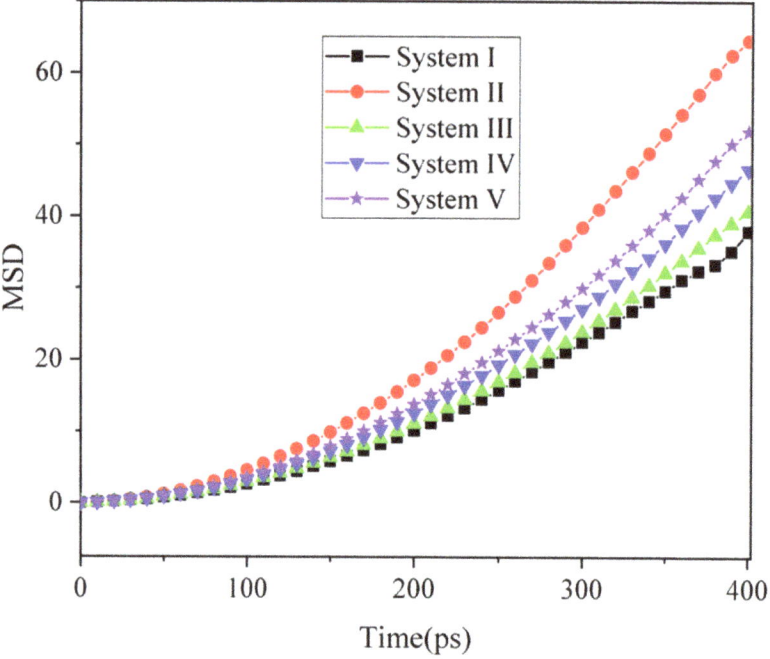

Figure 7. Mean square displacements of SDS and asphaltene molecules in five systems.

2.4. Aggregation Behavior of Oil Droplets

The purpose of electric field demulsification is to aggregate dispersed oil droplets to achieve oil/water separation. Conformations of the oil droplet at the beginning of the collision were selected as the initial structure to simulate the behavior of oil droplets after the shut-off of the electric field (see Figure 8). It can be seen that the oil droplets in contact with each other can continue to aggregate even in the absence of electric field. Taking System II as an example, we found that some asphaltene and SDS molecules, which guided the movement of the oil droplets, formed a contact surface between the two oil droplets and then migrated to the surface of the oil droplets under the influence of hydrophilic groups. At the same time, the hydrophobic components inside the interfacing oil droplets aggregated into a whole. Meanwhile, we calculated the radius of gyration (Rg) during the aggregation of oil droplets in the five systems. (as shown in Figure S2). We found that the radius of gyration of the oil droplets gradually decreased. Therefore, the droplets that collided would gradually aggregate into a whole.

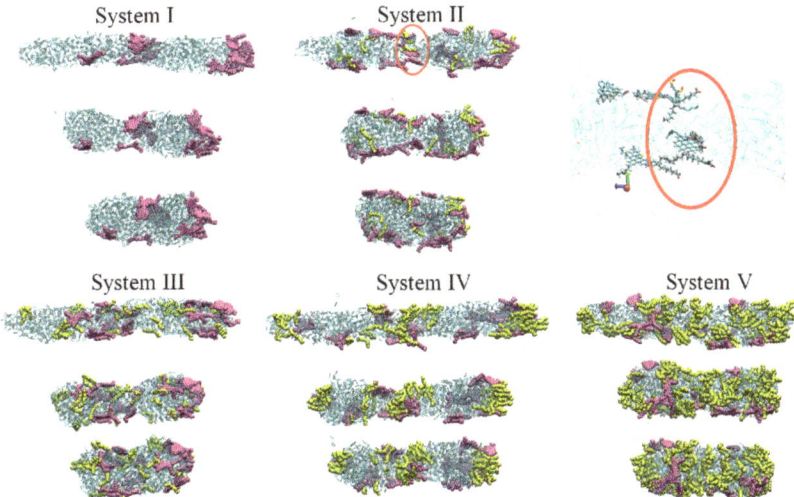

Figure 8. The change in the behaviors of oil droplets during the electric field shut-off period in five systems.

2.5. Mechanism of Aggregation of Oil Droplets

It had been proposed [35] that the surface charges of oil droplets will rearrange under BPEF. The positive and negative charges at the adjacent areas of the two oil droplets are opposite under the action of the electric field. Thus, the adjacent areas of oil droplets always attract each other along the BPEF direction. Here, we verified and explained the accumulation mechanism of oil droplets through theoretical methods. We calculated the interaction energy of the two oil droplets in all systems with E = 0.50 V/nm during the whole process from dispersion to aggregation. The calculated results are shown in Figure 9. The potential energy of the interaction between the two oil droplets is divided into two parts. The cyan areas represented the change in the potential energy between the oil droplets from dispersion to collision (i.e., electric field output durations), and the blue areas represented the change in the potential energy with time from collision to aggregation (i.e., electric field shut-off durations). At the same time, we calculated the root-mean-square deviation (RMSD) of crude oil droplets in the five systems (Figure S3). It can be seen from the figure that the aggregated oil droplets were basically stable after 4.0 ns. We found that the potential energy of the electrostatic interaction between the oil droplets was almost 0 kJ/mol during the entire electric field application process. The potential energy of the van der Waals interactions between the two oil droplets from dispersion to collision was also almost 0 kJ/mol in the output electric field stage. However, the potential energy of the van der Waals interactions between the oil droplets noticeably decreased after collision during the aggregation process (electric field shut-off stage). It means that in the electric field demulsification process, the adjacent areas of the two oil droplets with opposite charges have no obvious effect on the attraction and aggregation of the oil droplets. The van der Waals forces between the oil droplets are the main force in the demulsification process.

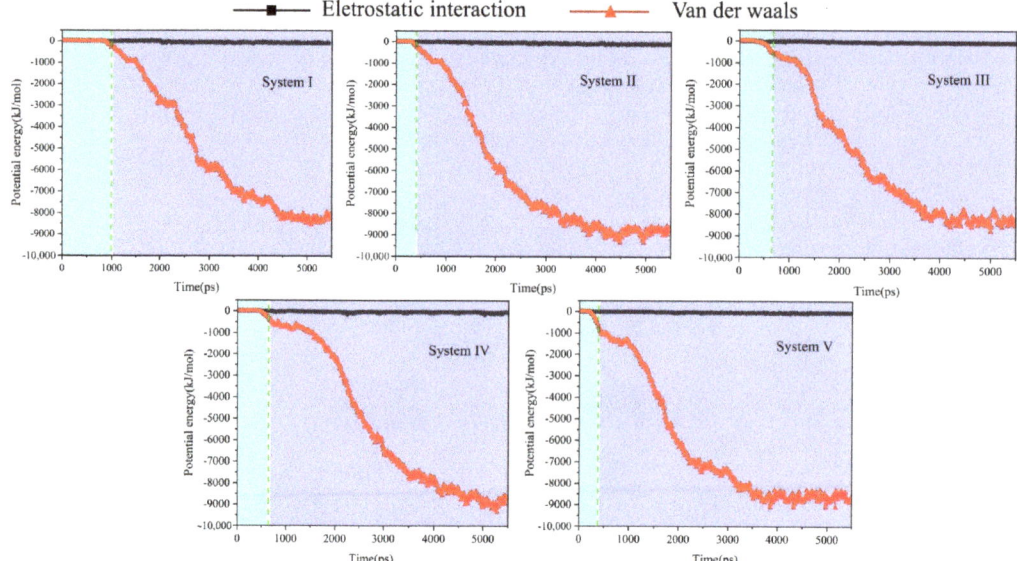

Figure 9. Interaction energy and its decomposed composition between two oil droplets during simulation.

3. Methods and Materials

3.1. Simulation Details

All MD simulations were performed in GROMACS 2019.6 software package. The GROMOS 53a6 force field [36] was used. The force field parameters of oil droplet composition were generated by the Automated Topology Builder (ATB) [37,38]. The simple point charge (SPC) model was selected for water molecules. The parameters of sodium ions (Na^+) that neutralize negative charge have been discussed in the literature [39].

Each system was energy-minimized using the steepest descent method before the simulation. The NVT ensemble at 300 K was performed with velocity rescaling thermostat. The NPT ensemble at 0.1 MPa and 300 K was performed with Berendsen pressure coupling. In the simulation, velocity rescaling thermostat with a time constant of 0.1 ps was selected as the temperature coupling method, and Berendsen pressure coupling with a time constant of 1.0 ps was selected as the pressure coupling method; the isothermal compression factor was set to 4.5×10^{-5} bar^{-1}. The periodic boundary condition was applied along three dimensions. During the simulation, van der Waals interactions used Lennard−Jones 12-6 potential, and the cutoff was set to 1.4 nm. The Coulombic interaction used particle-mesh Ewald (PME) summation method. The initial velocities were assigned according to Maxwell−Boltzmann distribution. The time step chose 2 fs. The trajectory was saved every 10 ps. VMD 1.9.3 was used for trajectory visualization.

3.2. Simulation Systems

3.2.1. Molecular Models of Crude Oil

Owing to the high complexity of crude oil, especially for asphaltene and resins, the asphaltenes demonstrate a key role in the stabilization of water-in-crude oil emulsions and significantly impact the rheological properties of crude oil [40]. Two types of asphaltene (i.e., the number of each type of asphaltene is four) and six types of resins [41] (i.e., the number of each type of resin is five) were selected based on previous studies, as shown in Figure 10. In addition to asphaltenes and resin molecules, four types of alkanes (32 hexane, 29 heptane, 34 octane and 40 nonane molecules), two types of cyclanes (22 cyclohexane and 35 cycloheptane molecules) and two types of aromatics (13 benzene and 35 toluene molecules) were selected as light oil components, referring to Song and Miranda's work [41–43]. Moreover, the concentration of resins and asphaltene in the crude oil was about 38%, which met the content of heavy oil components in crude oil [41].

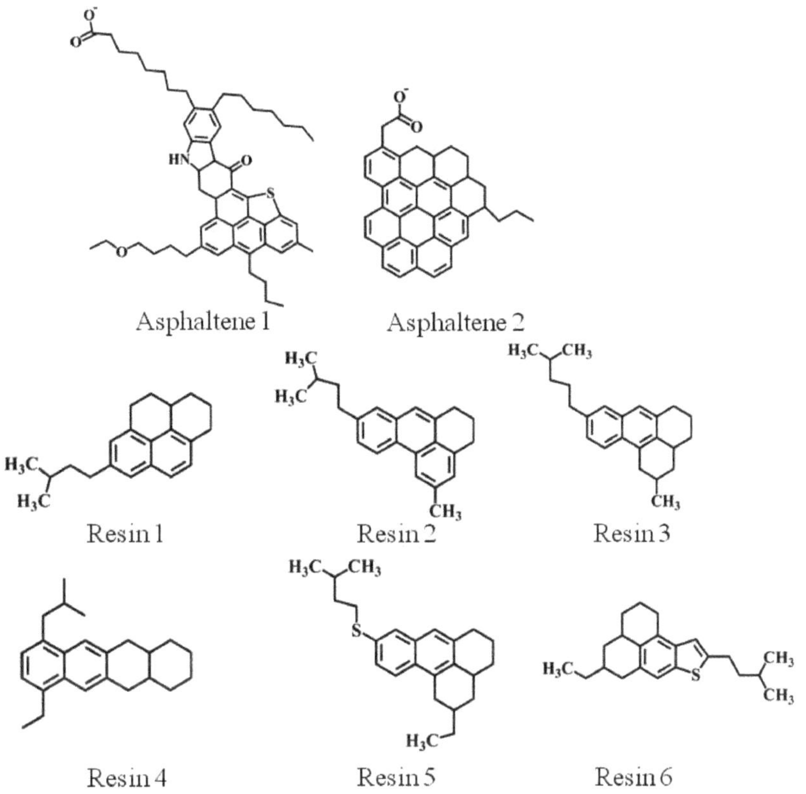

Figure 10. Asphaltene and resin molecules used in simulation.

3.2.2. Emulsified Oil Droplet

First, the components of crude oil including alkanes, cyclanes, aromatics, asphaltenes and resins were randomly inserted into a cubic box (x = 10 nm, y = 10 nm, z = 10 nm). To eliminate overlapping, energy minimization was then performed. After that, a 30 ns NPT ensemble simulation was performed to obtain a reasonable density. The equilibrium configuration after NPT run was shown in Figure 11a.

Second, the above crude oil was then solvated in an 8 nm × 8 nm × 8 nm simulation box with 19,230 water molecules. Energy minimization and a 20 ns NVT ensemble MD simulation was carried out to obtain the emulsified oil droplet (Figure 11b).

Third, emulsified oil droplets with different amounts of SDS adsorbed on their surface was constructed. SDS micelles were constructed using Packmol. The above spherical oil droplets were then placed in the center of a new box (10 nm × 10 nm × 15 nm) and SDS micelles were placed close to oil droplets (Figure 11c). Then, Na^+ counter ions and solvent were added. After energy minimization and a 20 ns NVT simulation, emulsified oil droplet systems were derived (Figure 11d).

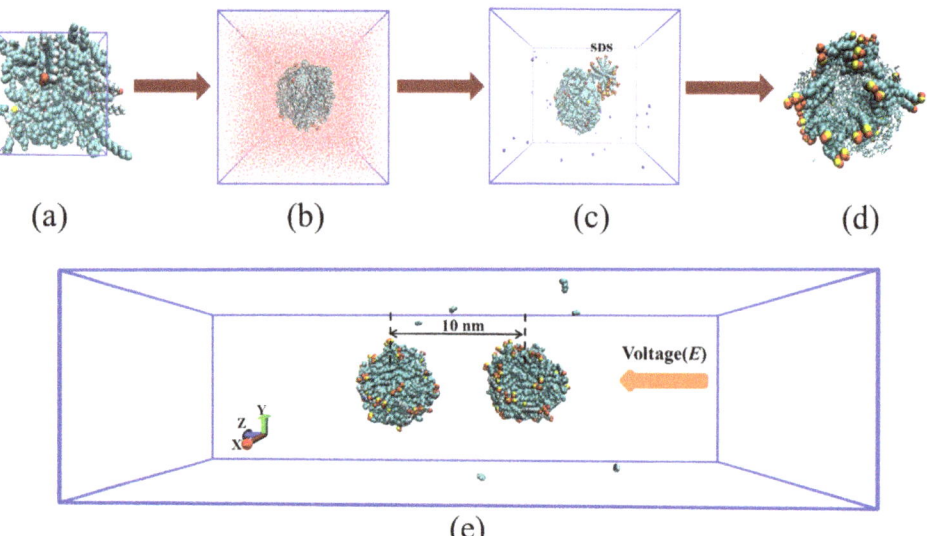

Figure 11. Schematic diagram of the building of emulsified oil droplet. (**a**) Equilibrium configuration of bulk crude oil after NPT simulation, (**b**) emulsified oil droplet in water, (**c**) initial configuration of emulsified oil droplet and SDS micelle, (**d**) Equilibrium configuration of emulsified oil droplet with SDS. For clarity, water molecules in (**c**,**d**) were not shown. (**e**) Lateral view of model simulation for the movement and aggregation behavior of oil droplets in O/W emulsion under BPEF. For clarity, water molecules in the system were not shown.

We assumed that the oil droplets distributed in the emulsion had the following conditions: First, the centroids of the two oil droplets were approximately along the z-axis direction; Second, the centroids of the two drops were about 10.0 nm apart; Finally, two identical emulsified oil drop models with counter ions were placed in a $10 \times 10 \times 50$ nm^3 box with a separation distance of about 10 nm, as shown in Figure 11e. Afterward, water molecules were added to solvate the system. The energy minimization and a 10 ns NVT simulation were applied to ensure emulsion system equilibrium. Subsequently, BPEF was imposed on all systems to study coalescence of the two droplets. The composition of each emulsified oil droplet system was shown in Table 1.

Table 1. Details of the emulsified oil droplet systems.

System	Number of Molecules				Mass Fraction (SDS of Oil Droplet)
	Crude Oil Droplet	SDS	Na$^+$	Water	
I	1	0	8	47,086	0.0%
II	1	10	18	47,018	6.2%
III	1	15	23	46,985	9.1%
IV	1	30	38	46,883	16.6%
V	1	60	68	46,680	28.5%

4. Conclusions

In this paper, molecular dynamics simulations were performed to study the behavior of oil droplets in O/W emulsion. The differences in oil droplets emulsified by different amounts of SDS were compared. Three major conclusions were derived. First, the hydrophilicity of oil droplets increases with increasing SDS content in the oil droplet. When electric field is applied, oil droplets move in the opposite direction of the electric field. The molecules in the oil droplet underwent redistribution. SDS and asphaltene with negatively charged functional groups were transferred to the head of the droplet along the direction of movement. The electrostatic potential surface of the oil droplet proved that the BPEF made the molecules redistribute in the droplet, which resulted in its surface potential redistribution as well. This is consistent with the theoretical hypothesis proposed by this experiment. Meanwhile, the collision time of oil droplets in all simulation systems was different due to the different SDS mass fraction, and the collision time was the shortest for the oil droplets with 6.2% SDS. The average elongation length le of the two oil droplets along the z direction explained that SDS molecules could change the elongation length of the oil droplets in the electric field. The MSD of SDS and asphaltene molecules under electric field showed that the mobility was the strongest in System II. Therefore, the elongation length of the oil droplets in System II was the largest, and this system was the least time consuming. Second, the oil droplets after collision can self-aggregate after electric field shut-off. SDS and asphaltene molecules on the contact surface between the two oil droplets migrated to the surface of the oil droplets under the influence of hydrophilic groups. Lastly, the adjacent areas of the two oil droplets with opposite charges have no obvious effect on the attraction and aggregation of the oil droplets, and the van der Waals forces between oil droplets are the main force in the demulsification process.

Supplementary Materials: The following supporting information can be downloaded at: https://www.mdpi.com/article/10.3390/molecules27082559/s1, Figure S1: Root-mean-square fluctuation (RMSF) of oil droplets in system I, system II and system IV, Figure S2: Radius of gyration (Rg) of oil droplets in five systems, Figure S3: Root-mean-square deviation (RMSD) of oil droplets in five systems.

Author Contributions: S.L. did the molecular dynamic simulations and wrote the manuscript. S.Y. supervised the molecular dynamic simulations and revised the manuscript. H.Z. designed and carried out the MD simulations and revised the manuscript. All authors have read and agreed to the published version of the manuscript.

Funding: This research was funded by Natural Science Foundation of Shandong Province (No. ZR2021MB040).

Institutional Review Board Statement: Not applicable.

Informed Consent Statement: Not applicable.

Data Availability Statement: Not applicable.

Acknowledgments: We gratefully appreciate the financial support from the Natural Science Foundation of Shandong Province.

Conflicts of Interest: The authors declare no conflict of interest.

Sample Availability: Not applicable.

References

1. Ahmadun, F.L.R.; Pen Da Shteh, A.; Abdullah, L.C.; Biak, D.; Madaeni, S.S.; Abidin, Z.Z. Review of technologies for oil and gas produced water treatment. *J. Hazard. Mater.* **2009**, *170*, 530–551.
2. Wahi, R.; Chuah, L.A.; Choong, T.; Ngaini, Z.; Nourouzi, M.M. Oil removal from aqueous state by natural fibrous sorbent: An overview. *Sep. Purif. Technol.* **2013**, *113*, 51–63. [CrossRef]
3. Zhu, Y.; Wang, D.; Jiang, L.; Jin, J. Recent progress in developing advanced membranes for emulsified oil/water separation. *NPG Asia Mater.* **2014**, *6*, e101. [CrossRef]
4. Abdullah, M.S.; Hilal, N.; Murali, R.S.; Padaki, M. Membrane technology enhancement in oil-water separation. A review. *Desalin. Int. J. Sci. Technol. Desalt. Water Purif.* **2015**, *357*, 197–207.
5. Zolfaghari, R.; Fakhru'L-Razi, A.; Abdullah, L.C.; Elnashaie, S.; Pendashteh, A. Demulsification techniques of water-in-oil and oil-in-water emulsions in petroleum industry. *Sep. Purif. Technol.* **2016**, *170*, 377–407. [CrossRef]
6. Srinivasan, A.; Viraraghavan, T. Removal of oil by walnut shell media. *Bioresour. Technol.* **2008**, *99*, 8217–8220. [CrossRef]
7. Ahmad, A.L.; Chong, M.F.; Bhatia, S.; Ismail, S. Drinking water reclamation from palm oil mill effluent (POME) using membrane technology—ScienceDirect. *Desalination* **2006**, *191*, 35–44. [CrossRef]

8. Simonović, B.R.; Dragana, A.; Mića, J.; Branimir, K.; Aca, J. Removal of mineral oil and wastewater pollutants using hard coal. *Chem. Ind. Chem. Eng. Q.* **2009**, *15*, 57–62. [CrossRef]
9. Uzunov, I.; Uzunova, S.; Gigova, A.; Minchev, L. Kinetics of oil and oil products adsorption by carbonized rice husks. *Chem. Eng. J.* **2011**, *172*, 306–311.
10. Annunciado, T.R.; Sydenstricker, T.; Amico, S.C. Experimental investigation of various vegetable fibers as sorbent materials for oil spills. *Mar. Pollut. Bull.* **2005**, *50*, 1340–1346. [CrossRef]
11. Lu, H.; Pan, Z.; Miao, Z.; Xu, X.; Wu, S.; Liu, Y.; Wang, H.; Yang, Q. Combination of electric field and medium coalescence for enhanced demulsification of oil-in-water emulsion. *Chem. Eng. J. Adv.* **2021**, *6*, 100103. [CrossRef]
12. Yang, S.; Sun, J.; Wu, K.; Hu, C. Enhanced oil droplet aggregation and demulsification by increasing electric field in electrocoagulation. *Chemosphere* **2021**, *283*, 131123. [CrossRef] [PubMed]
13. Sjblom, J.; Aske, N.; Auflem, I.H.; Brandal, Y.; Kallevik, H. Our current understanding of water-in-crude oil emulsions.: Recent characterization techniques and high pressure performance. *Adv. Colloid Interface Sci.* **2002**, *100*, 399–473.
14. Mhatre, S.; Vivacqua, V.; Ghadiri, M.; Abdullah, A.M.; Al-Marri, M.J.; Hassanpour, A.; Hewakandamby, B.; Azzopardi, B.; Kermani, B. Electrostatic phase separation: A review. *Chem. Eng. Res. Des.* **2015**, *96*, 177–195. [CrossRef]
15. Less, S.; Vilagines, R. The electrocoalescers' technology: Advances, strengths and limitations for crude oil separation. *J. Pet. Sci. Eng.* **2012**, *81*, 57–63. [CrossRef]
16. Zhao, Z.; Kang, Y.; Wu, S.; Sheng, K. Demulsification performance of oil-in-water emulsion in bidirectional pulsed electric field with starlike electrodes arrangement. *J. Dispers. Sci. Technol.* **2021**, 1–12. [CrossRef]
17. Hu, J.; Chen, J.; Zhang, X.; Xiao, J.; An, S.; Luan, Z.; Liu, F.; Zhang, B. Dynamic demulsification of oil-in-water emulsions with electrocoalescence: Diameter distribution of oil droplets. *Sep. Purif. Technol.* **2021**, *254*, 117631. [CrossRef]
18. Ptasinski, K.J.; Kerkhof, P.J.A.M. Electric Field Driven Separations: Phenomena and Applications. *Sep. Sci. Technol.* **1992**, *27*, 995–1021. [CrossRef]
19. Mousavichoubeh, M.; Ghadiri, M.; Shariaty-Niassar, M. Electro-coalescence of an aqueous droplet at an oil–water interface. *Chem. Eng. Process. Process Intensif.* **2011**, *50*, 338–344. [CrossRef]
20. Wang, S.S.; Lee, C.J.; Chan, C.C. Demulsification of Water-in-Oil Emulsions by Use of a High Voltage ac Field. *Sep. Sci.* **1994**, *29*, 159–170. [CrossRef]
21. Kim, Y.H.; Wasan, D.T.; Breen, P.J. A study of dynamic interfacial mechanisms for demulsification of water-in-oil emulsions. *Colloids Surf. A Physicochem. Eng. Asp.* **1995**, *95*, 235–247. [CrossRef]
22. Berg, G.; Lundgaard, L.E.; Abi-Chebel, N. Electrically stressed water drops in oil. *Chem. Eng. Process. Process Intensif.* **2010**, *12*, 1229–1240. [CrossRef]
23. Ichikawa, T.; Nakajima, Y. Rapid demulsification of dense oil-in-water emulsion by low external electric field.: II. Theory. *Colloids Surf. A Physicochem. Eng. Asp.* **2004**, *242*, 27–37. [CrossRef]
24. Hosseini, M.; Shahavi, M.H.; Yakhkeshi, A. AC and DC-Currents for Separation of Nano-Particles by External Electric Field. *Asian J. Chem.* **2012**, *24*, 181–184.
25. Bailes, P.J.; Freestone, D.; Sams, G.W. Pulsed DC fields for electrostatic coalescence of water-in-oil emulsions. *Chem. Eng.* **1997**, *38*, 34–39.
26. Bailes, P.J.; Larkai, S.K.L. An experimental investigation into the use of high voltage D.C. fields for liquid phase separation. *Trans. Inst. Chem. Eng.* **1981**, *59*, 229–237.
27. Ren, B.; Kang, Y. Aggregation of oil droplets and demulsification performance of oil-in-water emulsion in bidirectional pulsed electric field—ScienceDirect. *Sep. Purif. Technol.* **2019**, *211*, 958–965. [CrossRef]
28. Chen, Q.C.; Ma, J.; Xu, H.M.; Zhang, Y.J. The impact of ionic concentration on electrocoalescence of the nanodroplet driven by dieletrophoresis. *J. Mol. Liq.* **2019**, *290*, 111214. [CrossRef]
29. Chen, X.; Hou, L.; Li, W.; Li, S. Influence of electric field on the viscosity of waxy crude oil and micro property of paraffin: A molecular dynamics simulation study. *J. Mol. Liq.* **2018**, *272*, 973–981. [CrossRef]
30. He, X.; Wang, S.L.; Yang, Y.R.; Wang, X.D.; Chen, J.Q. Electro-coalescence of two charged droplets under pulsed direct current electric fields with various waveforms: A molecular dynamics study. *J. Mol. Liq.* **2020**, *312*, 113429. [CrossRef]
31. Ebrahimi, A.; Tamnanloo, J.; Mousavi, S.H.; Miandoab, E.S.; Hosseini, E.; Ghasemi, H.; Mozaffari, S. Discrete-Continuous Genetic Algorithm for Designing a Mixed Refrigerant Cryogenic Process. *Ind. Eng. Chem. Res.* **2021**, *60*, 7700–7713. [CrossRef]
32. Ghasemi, H.; Darjani, S.; Mazloomi, H.; Mozaffari, S. Preparation of stable multiple emulsions using food-grade emulsifiers: Evaluating the effects of emulsifier concentration, W/O phase ratio, and emulsification process. *SN Appl. Sci.* **2020**, *2*, 1–9. [CrossRef]
33. Wang, X.; Zhang, R.; Mozaffari, A.; Pablo, J.J.; Abbott, N. Active motion of multiphase oil droplets: Emergent dynamics of squirmers with evolving internal structure. *J. Mol. Liq.* **2020**, *312*, 113429. [CrossRef]
34. Farooq, U.; Lædre, S.; Gawel, K. Review of Asphaltenes in an Electric Field. *Energy Fuels* **2021**, *35*, 7285–7304. [CrossRef]
35. Ren, B.; Kang, Y. Demulsification of Oil-in-water (O/W) Emulsion in Bidirectional Pulsed Electric Field. *Langmuir* **2018**, *34*, 8923–8931. [CrossRef] [PubMed]
36. Oostenbrink, C.; Villa, A.; Mark, A.E.; Gunsteren, W. A biomolecular force field based on the free enthalpy of hydration and solvation: The GROMOS force-field parameter sets 53A5 and 53A6. *J. Comput. Chem.* **2010**, *25*, 1656–1676. [CrossRef]

37. Malde, A.K.; Zuo, L.; Breeze, M.; Stroet, M.; Poger, D.; Nair, P.C.; Oostenbrink, C.; Mark, A.E. An Automated Force Field Topology Builder (ATB) and Repository: Version 1.0. *J. Chem. Theory Comput.* **2011**, *7*, 4026. [CrossRef]
38. Koziara, K.B.; Stroet, M.; De Mal, A.K.; Mark, A. Testing and validation of the Automated Topology Builder (ATB) version 2.0: Prediction of hydration free enthalpies. *J. Comput.-Aided Mol. Des.* **2014**, *28*, 221–233. [CrossRef]
39. Wander, M.; Shuford, K.L. Molecular Dynamics Study of Interfacial Confinement Effects of Aqueous NaCl Brines in Nanoporous Carbon. *J. Phys. Chem. C* **2010**, *114*, 20539–20546. [CrossRef]
40. Mozaffari, S.; Ghasemi, H.; Tchoukov, P.; Czarnecki, J.; Nazemifard, N. Lab-on-a-Chip Systems in Asphaltene Characterization: A Review of Recent Advances. *Energy Fuels* **2021**, *35*, 9080–9101. [CrossRef]
41. Song, S.; Zhang, H.; Sun, L.; Sho, J.; Cao, X.; Yuan, S. Molecular Dynamics Study on Aggregating Behavior of Asphaltene and Resin in Emulsified Heavy Oil Droplets with Sodium Dodecyl Sulfate. *Energy Fuels* **2018**, *32*, 12383–12393. [CrossRef]
42. De Lara, L.S.; Michelon, M.F.; Miranda, C.R. Molecular Dynamics Studies of Fluid/Oil Interfaces for Improved Oil Recovery Processes. *J. Phys. Chem. B* **2012**, *116*, 14667–14676. [CrossRef] [PubMed]
43. Kunieda, M.; Nakaoka, K.; Liang, Y.; Miranda, C.R.; Ueda, A.; Takahashi, S.; Okabe, H.; Matsuoka, T. Self-accumulation of aromatics at the oil-water interface through weak hydrogen bonding. *J. Am. Chem. Soc.* **2010**, *132*, 18281–18286. [CrossRef] [PubMed]

Article

Determination of Minimum Miscibility Pressure of CO_2–Oil System: A Molecular Dynamics Study

Ding Li [1,2], Shuixiang Xie [1], Xiangliang Li [3], Yinghua Zhang [3], Heng Zhang [2,*] and Shiling Yuan [2]

1. State Key Laboratory of Petroleum Pollution Control, CNPC Research Institute of Safety & Environment Technology, Beijing 100000, China; blyceoh@163.com (D.L.); tu_95ms@sina.com (S.X.)
2. Key Lab of Colloid and Interface Chemistry, Shandong University, Jinan 250100, China; shilingyuan@sdu.edu.cn
3. Shengli Oil Field Exploration and Development Research Institute, Dongying 257000, China; lxliang1964@163.com (X.L.); zhangyh3000@foxmail.com (Y.Z.)
* Correspondence: zhangheng@sdu.edu.cn

Abstract: CO_2 enhanced oil recovery (CO_2-EOR) has become significantly crucial to the petroleum industry, in particular, CO_2 miscible flooding can greatly improve the efficiency of EOR. Minimum miscibility pressure (MMP) is a vital factor affecting CO_2 flooding, which determines the yield and economic benefit of oil recovery. Therefore, it is important to predict this property for a successful field development plan. In this study, a novel model based on molecular dynamics to determine MMP was developed. The model characterized a miscible state by calculating the ratio of CO_2 and crude oil atoms that pass through the initial interface. The whole process was not affected by other external objective factors. We compared our model with several famous empirical correlations, and obtained satisfactory results—the relative errors were 8.53% and 13.71% for the two equations derived from our model. Furthermore, we found the MMPs predicted by different reference materials (i.e., CO_2/crude oil) were approximately linear ($R^2 = 0.955$). We also confirmed the linear relationship between MMP and reservoir temperature (T_R). The correlation coefficient was about 0.15 MPa/K in the present study.

Keywords: minimum miscible pressure; CO_2 enhanced oil recovery; molecular dynamics

1. Introduction

Global warming has caused great changes such as continued sea level rise, which is irreversible over hundreds to thousands of years. CO_2 is the culprit of this phenomenon. CCUS (CO_2 capture, utilization, and storage) is a new technology developed from CCS (CO_2 capture and storage) that can bring economic benefits while reducing CO_2 emissions and alleviating global warming [1]. CO_2 enhanced oil recovery (CO_2-EOR) is one of the effective ways of CCUS. The captured CO_2 is squeezed into the oil reservoirs that have been exploited, and the interaction between CO_2 and crude oil is used to improve the properties of the crude oil, thereby displacing more crude oil from the crust [2]. Research has shown that CO_2-EOR can improve crude oil recovery significantly and extend the life of oil reservoirs [3,4]. Hence, CO_2-EOR has been fundamentally well researched in laboratories and applied in industries as an efficient approach since the 1970s [5].

There are two different miscible and immiscible states in CO_2-EOR. Under the former condition, CO_2 and crude oil can completely integrate into one phase, resulting in a much higher recovery rate than the latter. For the former, there is a minimum pressure above which CO_2 and crude oil can be miscible. This minimum pressure value, also called the minimum miscible pressure (MMP), is a vital parameter in the process of CO_2-EOR. Nevertheless, considering the massive influencing factors, the accurate determination of MMP remains a major challenge [6].

To date, there are various ways to predict MMP such as experimental measurement and computational methods. The former has been widely used due to their high precision. Within them, slim-tube experiments [7–9], as a necessary test in the industry, is considered to be the standard experimental procedure. Rising-bubble apparatus (RBA) [10,11] and vanishing interfacial tension (VIT) [12–15] are also frequently utilized to determine MMP because of their simplicity and flexibility. Although these experimental measurements have accurate techniques, they still suffer from some disadvantages including time-consumption and operation cost. Furthermore, it is difficult for any experimental method to simulate the real conditions of the crude oil reservoirs completely so that their results are greatly influenced by the instruments.

The application of computational techniques is an available alternative approach to experiments. In 1960, the first empirical MMP correlation was proposed by Benham et al. [16]. The reported equation was correlated using three pseudo-components presenting a multi-components system, and some satisfactory results were obtained based on this model. Thereafter, an increasing number of correlations were developed for MMP prediction [17–20]. Researchers found that the more useful parameters an equation used, the better performance the model had [21]. These parameters generally included reservoir temperature (T_R), composition of drive gas (CO_2, H_2S, N_2, and C_1–C_5), molecular weight of C_{5+} fraction in crude oil (MW_{c5+}), and the ratio of volatile (C_1 and N_2) to intermediate (C_2–C_4, H_2S, and CO_2) in crude oil (Vol./Int.).

In addition to the conventional empirical formula models, the parameters above are often used in some intelligent algorithms based on machine learning. For instance, artificial neural networks (ANNs) can learn from large amounts of input data, and reflect their relationships more effective than conventional techniques [22]. Determination of network structure and its parameters are two crucial steps in achieving high performance from ANN. One part of the data is used to train and look for a suitable structure and optimal parameters, while the other tests the prediction accuracy of the model. Based on the principle, back propagation (BP) [23] and radial basis function (RBF) [24] are proposed. Beyond that, a series of optimization methods such as genetic algorithm (GA) [25], particle swarm optimization (PSO) [26], support vector machine (SVM) [27], and hybrid-ANFIS [28] have also been developed for MMP determination. In a previous study [29], we compared four estimation methods and found that the machine learning intelligent algorithm had a higher precision to the MMP than pure linear model. In addition, some reports that combined multiple approaches showed better results [30–34].

However, all of the above methods cannot give a direct explanation of the MMP from a microscopic view. They are all based on the existing oilfield data, which means that the established model will inevitably be affected by specific situation. To put it another way, these methods can be considered as pure mathematical statistics methods that have low levels of universality for different CO_2-EOR.

Against this backdrop, the current study proposes a novel MMP prediction model at the molecular level, and the research process was not affected by other external objective factors. Therefore, the model represents a new strategy. First, we built a simulation box that contained CO_2 and crude oil with an obvious phase interface. To mimic the contact between CO_2 and crude oil, these molecules were gradually mixed until they were miscible with time evolution by using molecular dynamics. After calculating the ratio of CO_2/crude oil atoms that passed through the initial interface, we found the connection between the ratio value with the miscible state. When the ratio changes from decreasing to stable, it indicates that the system has entered a miscible state, while the pressure corresponding to the inflection point is MMP. Figure 1 is a flow chart that shows all the main steps of modeling. The main objective of this study was to reveal the principle of the MMP formation at the molecular level and provide more feasible ideas for the prediction of the MMP.

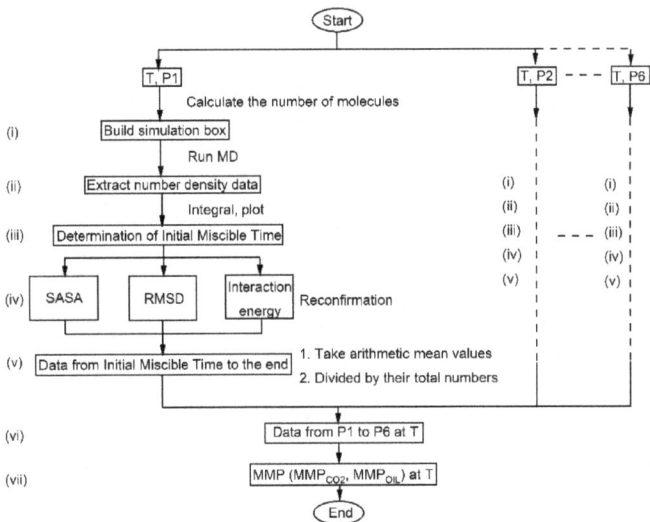

Figure 1. Flowchart of proposed MMP prediction model. (i) Construction of the simulation system, (ii) Extracting number density data after MD simulation, (iii) Determination of initial miscible time, (iv) Reconfirmation of initial miscible time, (v) Processing data from initial miscible time to the end, (vi) Processing data from P1 to P6 at T, (vii) Acquisition of MMP.

2. Simulation Method

2.1. Simulation and Force Field

The molecular dynamics simulation was performed by the GROMACS 4.6.7 package [35,36], and AMBER 03 all-atom force field [37]. Parameters set for all components of crude oil and CO_2 were generated from Automated Topology Builder and Repository databases [38,39].

The convergence criterion of energy minimization was 1000 kJ/(mol·nm). In the simulation, a velocity rescaling thermostat with a 0.1 ps time constant was selected as the temperature coupling method [40]. Berendsen pressure coupling with 1.0 ps time constant was selected as the pressure coupling method. The isothermal compression factor was set to 4.5×10^{-5} bar^{-1} [41]. The time step was 2 fs, and periodic boundary conditions were applied in the XY directions [42]. Walls were set at the top and bottom of the Z-direction in the simulated box to ensure that all atoms passed through the initial interface to achieve the miscibility. Bond lengths were constrained by the LINCS algorithm [43]. During the simulation, van der Waals interactions with the Lennard–Jones potential was cut off at 1.4 nm. Coulomb interaction used the particle-mesh Ewald summation method [44,45]. The Verlet list was updated every 10 steps. The Maxwell–Boltzmann distribution was employed to set the initial atomic velocities of the systems [46]. The trajectories were integrated by the leapfrog Verlet algorithm [47].

2.2. Simulation System

In a real situation, the chemical components of crude oil are highly complex. Under the current experimental conditions, it is time-consuming to precisely analyze the exact constitution of its components. In order to get as close as possible to the real situation, the oil model was designed based on Miranda's works [48,49], which were used to explore the interface properties between crude oil and different fluids. Their model contained alkanes (72 hexane, 66 heptane, 78 octane, and 90 nonane molecules), cyclanes (48 cyclohexane and 78 cycloheptane molecules), and aromatics (30 benzene and 78 toluene molecules), and has been proven reliable by Song et al. [50].

At 333 K and 10 MPa, all alkanes, cyclanes, and aromatics were added into a cubic box (x = 9 nm, y = 9 nm, z = 9 nm) randomly. Then, energy minimization was performed to eliminate opposed-conformation. In order to mimic the state of crude oil in the reservoir, we performed a 30 ns NPT ensemble simulation to obtain its equilibrium state. After equilibration, the size of simulation box changed to 5.2 nm × 5.2 nm × 5.2 nm.

Furthermore, we built a box of the same size, stochastically adding 561 CO_2 molecules to mimic the supercritical CO_2 fluid (333 K, 10 MPa). Energy minimization and 30 ns NVT ensemble simulation enabled the CO_2 to reach its equilibrium state. To simulate the contact between CO_2 and crude oil, the two boxes were integrated into one rectangular simulation box, and the height of new box in the Z-direction was slightly increased to 11.2 nm to avoid intermolecular overlap, as shown in Figure 2. After that, at least 10 ns NPT ensemble MD simulation was performed.

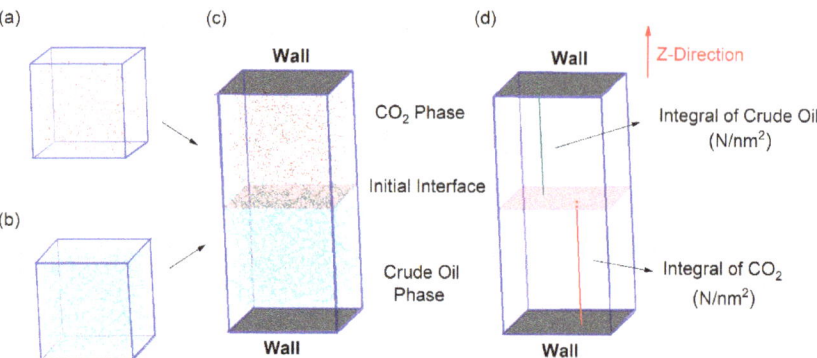

Figure 2. Construction of the simulation system. (**a**) CO_2 phase. (**b**) Crude oil phase. (**c**) Initial simulation box. (**d**) Integral of CO_2/crude oil molecules passing through the initial phase interface.

3. Results and Discussion

In the last NPT ensemble simulation, the size of the box gradually stabilized over time. When CO_2 was miscible with crude oil, the NPT ensemble achieved equilibrium and the box size remained unchanged. However, the change in the size of the box cannot intuitively reflect the miscibility process. Therefore, we introduced the number density of CO_2/crude oil.

The product of number density and volume of a box is the total number of atoms. When the system achieves equilibrium, the ratio of CO_2/crude oil atoms in the upper or lower half of the box should be 50%. In our simulation system, the Z-direction height will not change due to the presence of walls. Therefore, the integral change in the number density in the Z-direction (i.e., the integral bars in Figure 2d) can reflect the change in the size of the box and further reflect the mixing progress. When the system achieves equilibrium, the integral of the number density in Z-direction will also be constant.

3.1. Definition of Initial Miscible Time

First, the initial miscible time was defined. It refers to the moment when CO_2 and the crude oil phases just reach the miscible state during their mixing progress, and they can keep the miscible state afterward. The purpose is to ensure that data after this time are miscibility data. We used the CO_2 phase as an example to illustrate the calculations. Its number density data were extracted along the Z-direction of the box from 0.5 ns to 10.0 ns every 0.5 ns after the NPT ensemble was run. Hence, there were 20 sets of data in total. We can obtain the number of CO_2 atoms in the lower half of the box by integrating the density of CO_2 along the Z-direction below the initial interface at each cut-off time. Since CO_2 was not distributed in the box below the initial interface at the beginning, the integral values we obtained corresponded to the number of CO_2 atoms passing through the initial interface at the cut-off time. More vividly, it is an integral bar in the three-dimensional space of the box along the Z-direction, as shown in Figure 2d.

Furthermore, a curve of the number of CO_2 atoms passing through the initial interface over time can be plotted. Figure 3 shows the change in the number of CO_2 atoms in the lower half of the box at 333 K and 10 MPa: it gradually increased from zero to a stable value (about 49.09), and then tended to be stable. It is worth noting that each molecule always kept in continuous random motion, thus it is normal to have positive and negative fluctuations after miscibility. For the selection of the initial miscibility time, the establishment standard is to find the time when the curve becomes stable and the change is very gentle after the miscibility reference line (i.e., the 49.09 line in Figure 3a). This is also the time when the miscibility has just been achieved. The first-order variance of data with time evolution (Figure 3b) reflects the trend of data changes more intuitively. It should be noted that it is not the "initial miscibility" that is already zero, but the time corresponding to the point relatively close to zero. Based on the situation in Figures 2 and 3, it can be guaranteed that the time at 4 ns: (i) the vertical axis value is already very close to the reference line, and (ii) the curve's upward trend has slowed down.

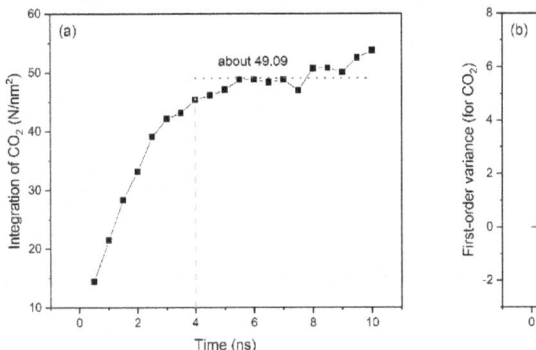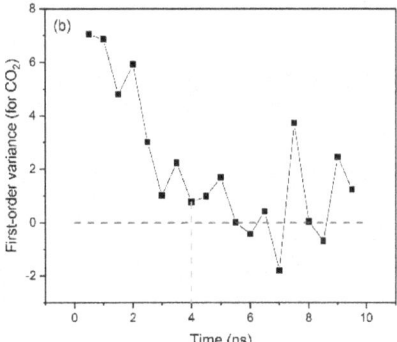

Figure 3. The number of CO_2 atoms passing through the initial interface (**a**) and its first–order variance (**b**) with time evolution (333 K, 10 MPa).

Based on similar treatments, crude oil atoms passing through the initial interface with time evolution (Figure S1) and its first-order variance (Figure S1b) can also be drawn. Figure 3 and Figure S1 show that the changes in CO_2 and crude oil were quite similar. Combined with the analyses above, we can preliminarily conclude that 4 ns is the initial miscible time, and is also the time when the CO_2–oil system achieved miscibility.

3.2. Reconfirmation of Initial Miscible Time

3.2.1. Solvent Accessible SURFACE Area (SASA) Analysis

To confirm the initial miscible time discussed in previous parts, solvent accessible surface area (SASA) was calculated. SASA represents the hydrophobic, hydrophilic, and total solvent accessible surface area for each component of the simulated system. Figure 4 shows the change in the hydrophilic area of CO_2 from 0 ns to 10 ns and went through roughly three processes: (i) At the beginning, SASA increased rapidly and reached the highest point (from C1 to C2); (ii) SASA dropped to the lowest point in a short period of time (from C2 to C3); and (iii) SASA gradually rose to basic stability and attained a state of dynamic balance (from C3 to C4, and the time was after about 4 ns).

C1 and C2 are adjustments to the initial configuration in the molecular dynamics NPT ensemble, which was not our focus. This can be contributed to the molecular dynamic method being able to readjust the molecular conformation in the model under the NPT ensemble, and we focused more on the change in conformation after being readjusted. With the blending of CO_2 and crude oil phases, both gradually achieved the best coexistence state (after C4).

Similarly, the changing trend of hydrophobic surface area of crude oil can be obtained by the same method. As shown in Figure 4b, a similar SASA change was observed in which the area increased and then decreased rapidly with time evolution (from O1 to O3 in Figure 4b). From Figure 4, it is reasonable to select 4 ns as the initial miscible time, and the data after 4 ns can be used to discuss the miscibility.

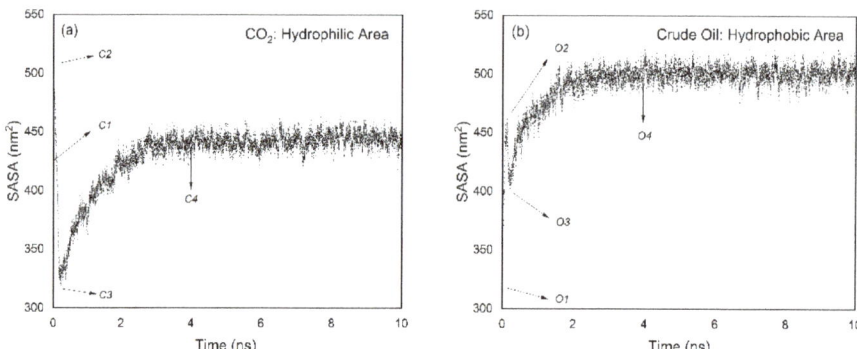

Figure 4. SASA analysis for CO_2 (**a**) and crude oil (**b**).

3.2.2. Root Mean Square Deviation (RMSD) Analysis

Root mean square deviation (RMSD) compares each molecular structure in the simulation from the trajectory to the initial reference structure, reflects the change in its conformation, and is calculated by Equation (1).

$$RMSD = \sqrt{\frac{1}{N}\sum_{i=1}^{N}(|r_i(t) - r_i(0)|)^2} \quad (1)$$

where N is the total number of atoms (CO_2/crude oil); and $r_i(0)$ and $r_i(t)$ are the initial position and the position of atom i at time t. Figure 5 displays the RMSD of CO_2/crude oil during NPT ensemble as a function of time. It is interesting to note that CO_2 has a higher RMSD value than crude oil at the beginning, which indicates that CO_2 has better mobility. From 4 ns to 10 ns, the RMSD of CO_2/crude oil in the box fluctuated with time evolution. Both were gathered around 4 nm of RMSD, which signifies that the system achieved equilibrium after 4 ns.

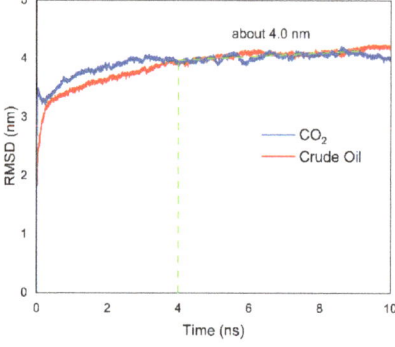

Figure 5. RMSD analysis for CO_2 and crude oil.

3.2.3. Interaction Energy Analysis

The energy changes can reveal the changes in conformation in the simulated system and represent the miscibility process between CO_2 and crude oil phases. Interaction energy is a type of non-bonding interaction including long-range Coulomb interaction and short-range van der Waals interaction. As shown in Figure 6, the system was dominated by van der Waals interaction, while Coulomb interaction accounts for only about one-tenth of the former. This is because both CO_2 and crude oil are non-polar molecules and do not have forces such as strong hydrogen bonding interaction. The intermolecular forces are mainly dispersive forces. The dispersion forces increased with time evolution, and the van der Waals potential energy and the total intermolecular potential energy increased accordingly.

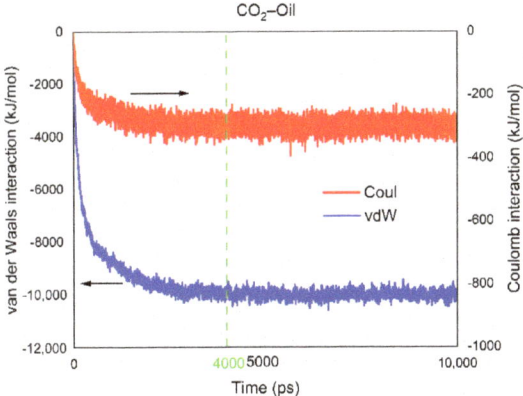

Figure 6. Interaction energy analysis.

When the system achieved equilibrium, the total interaction energy between CO_2 and crude oil also reached its maximum and remained dynamically stable. Figure 6 clearly indicates that van der Waals interaction and Coulomb interaction both remained stable after 4 ns.

3.3. Acquisition of MMP

Once the initial miscible time system at 333 K and 10 MPa has been successfully determined, the number of CO_2 atoms in the lower half of the box and crude oil atoms in the upper half of the box after 4 ns were taken as the arithmetic mean respectively. It needs to point out that the number of CO_2 molecules under different pressures are different for 333 K system (Table S1). In order to reflect the general laws, the ratio of mean value to their respective total number of CO_2 and crude oil atoms in the box was calculated. Similarly, the ratio of CO_2/crude oil passing through the initial interface to their respective totals at 333 K for 15 MPa, 20 MPa, 25 MPa, 30 MPa, and 35 MPa were also calculated, as shown in Table 1.

Table 1. The ratio of CO_2 and crude oil atoms passing through the initial interface to their respective totals.

	CO_2	Crude Oil
10 MPa	0.0292	0.0300
15 MPa	0.0231	0.0236
20 MPa	0.0219	0.0221
25 MPa	0.0216	0.0217
30 MPa	0.0204	0.0207
35 MPa	0.0205	0.0208

From 10 to 35 MPa, the data of ratio decreased first and then became stable. We believe that this is because the system reached its peak pressure at 333 K. When the system exceeded this pressure, the additional simulation will not affect the value of each ratio. Therefore, the pressure is the theoretical MMP at 333 K.

For the sake of confirming the MMP, we handled the data according to its regularity. The first three decreasing points were fitted linearly, representing the systems before MMP, and an equation in the form of y = kx + b was obtained. The last three nearly equal points were regarded as stable points, representing the systems after MMP, thus, another equation of y = x can be acquired by taking their arithmetic mean. We can subsequently obtain an intersection point as a consequence of simultaneous equations, and the abscissa corresponding to this point is the exact MMP at 333 K. As shown in Figure 7a,b, it was 20.31 MPa for CO_2 and 20.21 MPa for crude oil.

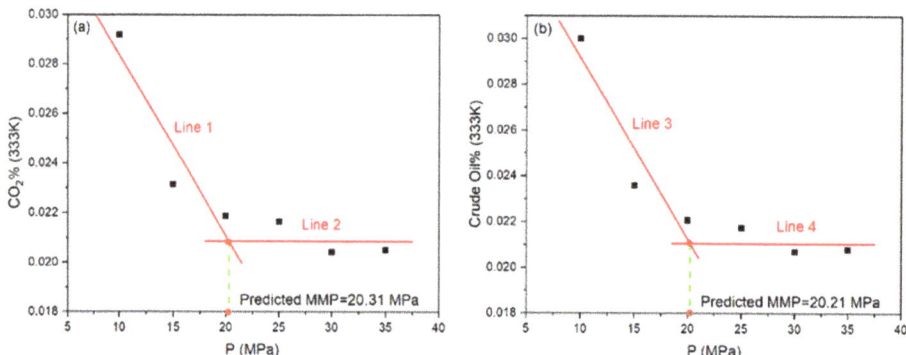

Figure 7. Acquisition of MMP (333 K) for CO_2 (**a**) and crude oil (**b**).

3.4. MMP in Different Temperature Systems

We continued to simulate and analyze the data under the condition of 343 K, 353 K, 363 K, and 373 K at 10 MPa, 15 MPa, 20 MPa, 25 MPa, 30 MPa, and 35 MPa, respectively. It is worth pointing out that the density of CO_2 varies greatly at different temperatures and pressures, therefore we computed the number of CO_2 molecules under different conditions. The amount of CO_2 molecules added to each simulation system are listed in Table S1. Afterward, a summary of initial miscible time in different systems can be obtained according to the methods in Sections 3.1 and 3.2, as listed in Table 2. Table S2 summarized all the integral values in this study.

Table 2. Summary of initial miscible time (ns) in different systems.

	333 K	343 K	353 K	363 K	373 K
10 MPa	4.0	5.0	3.5	4.0	3.0
15 MPa	4.5	6.0	3.5	3.0	2.5
20 MPa	4.5	5.0	4.0	3.0	3.0
25 MPa	4.0	3.5	4.0	3.0	3.5
30 MPa	3.5	4.0	4.0	3.0	4.0
35 MPa	4.5	4.0	4.5	3.5	4.0

The ratio of CO_2 and crude oil atom numbers that passed through the initial interface to their respective totals when they achieved miscibility can be obtained. Consequently, MMP of 343 K, 353 K, 363 K, and 373 K were obtained according to the method described in Section 3.4 by plotting and curve fitting (Figure S2). Table 3 summarizes the results.

Table 3. Summary of MMP (MPa) obtained from CO_2/crude oil in different systems.

	CO_2	Crude Oil
333 K	20.31	20.21
343 K	21.08	20.89
353 K	22.12	22.36
363 K	24.43	23.84
373 K	26.25	24.52

3.5. Model Assessment

We fitted the MMP obtained from CO_2/crude oil to T_R, respectively, and obtained two prediction equations (Figure 8). It can be compared with the experimental results to check the predictive performance of the model. Recently, Yu et al. used a combination method of slim-tube experiments and interfacial tension (IFT) to perform MMP measurements on tight oil from the Long Dong region of the Ordos Basin. This method has higher credibility than slim-tube experiments [51]. Afterward, we compared our model with several famous empirical correlations to illustrate its accuracy by employing the experimental method proposed by Yu et al. as the benchmark. Table 4 reports the relative error. Details of these empirical correlations are summarized in Table S3.

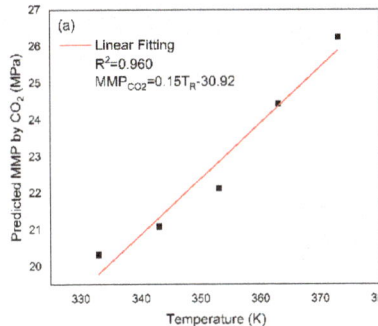
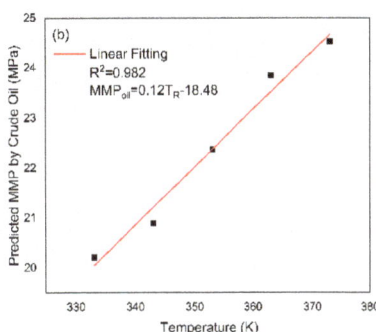

Figure 8. Relationships between T_R and MMP for CO_2 (**a**) and crude oil (**b**).

Table 4. Summary of MMP (MPa) and relative error predicted by experimental and different empirical correlations.

Model	Number of Parameters	Predicted MMP (MPa)	Relative Error (%)
Yu et al. [51]	-	22.75	-
CO_2 (this study)	1	19.63	13.71
Crude Oil (this study)	1	20.81	8.53
Lee [52]	1	20.84	8.32
Alston et al. [53]	4	19.72	13.22
Shokir [54]	8	20.03	11.89
Emera and Sarma [25]	2	30.11	32.44
Cronquist [55]	3	26.59	16.96
Glaso [56]	2	27.60	21.41
Yellig and Metcalfe [57]	1	16.55	27.18

The relative error obtained from crude oil was similar to Lee [52], and the CO_2 relative error was similar to Alston et al. [53]. The equation proposed by Shokir [54] was based on an alternating conditional expectation algorithm, and had a relative error of 11.89 %. The model of Emera and Sarma [25] can be employed to calculate the MMP of impure CO_2 injection, but has poor accuracy. Beyond that, the performances of Cronquist [55],

Glaso [56], and Yellig and Metcalfe [57] were also unsatisfactory. The overall results can prove that even if only the influencing factor of T_R is considered, the model proposed in this study had satisfactory prediction accuracy.

3.6. Comparison of MMP Predicted by CO_2 and Crude Oil

The relationships between MMP predicted by CO_2 and crude oil can be compared. It is more intuitive to reflect the data in Table 3 to Figure 9. In Figure 9, the blue line represents the curve whose analytical formula is y = x, and the red line is the fitting curve for the data. It can be found that the MMPs predicted by CO_2 and crude oil were approximately linear (R^2 = 0.955). Furthermore, in the same simulation system, the MMP values obtained from different reference materials (CO_2/crude oil) were not identical as there was a slight difference between them (i.e., an included angle of about 8°).

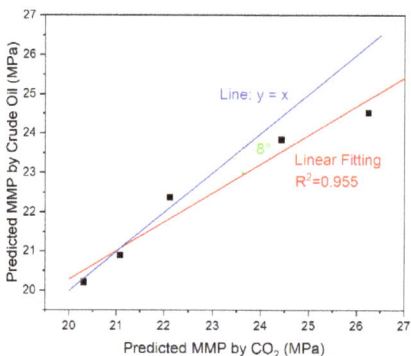

Figure 9. Comparison of MMP predicted by CO_2 and crude oil.

3.7. Relationships between T_R and MMP

However, the real situation of each oil reservoir varies, and the composition of injected gases is also different in EOR, so it is meaningless and almost impossible to obtain the accurate relationship between each influence factor and MMP. For a certain influencing factor, we can explore the qualitative relationship between the factor and MMP. Oil reservoir temperature (T_R) is usually regarded as one of the most important factors affecting MMP [58]. Exploring the influence of T_R on MMP is the core of many studies (such as the fitting of empirical formula). Recently, Zheng et al. [59] proposed a novel oil droplet volume measurement method (ODVM) to measure the multiple contact minimum miscibility pressure (MCMMP) and first contact miscibility pressure (FCMP) in the CO_2/n-hexadecane ($C_{16}H_{34}$) and CO_2/liquid paraffin systems. Their experimental data showed that the measured MMP values of two CO_2–oil systems increased linearly with T_R. Furthermore, Mostafa et al. found that the MMP is a linear function of temperature with a slope of 0.15 MPa/K [60].

The modeling method of this study shows that the relationship between T_R and MMP can be identified in the principle of miscibility because it is not affected by other external objective factors. As shown in Figure 8, for both CO_2 and crude oil, the change in T_R and MMP basically conformed to a linear relationship, thus a fairly good fitting result can be obtained by using the first-order linear equation. This is because the increase in T_R can effectively reduce the solubility of CO_2 in crude oil, which is not conducive to the mixing progress of CO_2 and crude oil, ultimately leading to the increase in MMP. During the temperature range (333–373 K), it is a linear change with a slope of 0.15 MPa/K and 0.12 MPa/K and consistent with the experimental results.

4. Conclusions

In this paper, a novel molecular dynamics-based model to determine minimum miscible pressure of CO_2–oil system was developed. The model characterized the miscible state by calculating the ratio of both CO_2 and crude oil atoms that passed through the initial interface to their respective totals. These ratio values dropped rapidly and fluctuated after a certain value with the increase in pressure at a fixed T_R. The value is the MMP of T_R. In comparison with conventional prediction approaches, the present work proposed a straightforward model to simulate the complex miscibility of CO_2 and crude oil, and the miscible principle was clarified at the molecular scale.

Based on the above studies, the newly proposed model is believed to be reliable for the prediction of MMP. However, there still remain some distinctions when compared to the real situation, which may have a certain impact on the prediction [61]. We have begun to adjust the model to enhance its application. For example, we plan to introduce silica slab and asphaltenes to mimic the real situation of crust and heavy oil, respectively. To sum up, the following conclusions can be drawn:

(1) The molecular scale mixing progress of CO_2 and crude oil was investigated in principle for the first time, and the research process was not affected by other external objective factors. Results showed that the ratio of CO_2/crude oil atoms that passed through the initial interface to their respective totals was always the same when the system was miscible. The proposed model had good prediction capabilities.

(2) In the process of the simulation, the SASA, RMSD, and interaction energy of CO_2/crude oil changed obviously, thus they can be used as criteria of miscibility between both phases.

(3) The MMP predicted by CO_2 of the CO_2–oil system were 20.31 MPa, 21.08 MPa, 22.12 MPa, 24.43 MPa, and 26.25 MPa at temperatures of 333 K, 343 K, 353 K, 363 K, and 373 K, respectively, and MMPs predicted by crude oil were 20.21 MPa, 20.89 MPa, 22.36 MPa, 23.84 MPa, and 24.52 MPa at the same temperatures. The two sets of data had a linear relationship.

(4) MMP and reservoir temperature (T_R) had a linear relationship in the present work, and the slope was about 0.15 MPa/K, which are in agreement with theoretical analyses and literature results.

Supplementary Materials: The following are available online. Figure S1: The number of crude oil atoms passing through the initial interface. Figure S2: Acquisition of MMP in different systems. Table S1: The number of CO_2 molecules added in different systems. Table S2: Integrated values of CO_2 and crude oil at different temperatures. Table S3: Summarization of some famous empirical correlations.

Author Contributions: Conceptualization, D.L. and H.Z.; methodology, D.L., H.Z. and S.Y.; software, D.L.; validation, D.L., H.Z. and S.Y.; formal analysis, D.L. and H.Z.; investigation, D.L.; resources, S.X., X.L., Y.Z. and S.Y.; data curation, D.L.; writing—original draft preparation, D.L.; writing—review and editing, D.L., H.Z. and S.Y.; visualization, D.L.; supervision, H.Z.; project administration, S.X., X.L. and Y.Z.; funding acquisition, S.Y. All authors have read and agreed to the published version of the manuscript.

Funding: This research project was funded by the financial support from the National Science Foundation of China (No. 21573130) and the Youth Innovation Group of Shandong University (No. 2020QNQT018).

Institutional Review Board Statement: Not applicable.

Informed Consent Statement: Not applicable.

Data Availability Statement: The data presented in this study is available upon reasonable request.

Conflicts of Interest: The authors declare no conflict of interest.

Sample Availability: Samples of the compounds are not available from the authors.

References

1. Farajzadeh, R.; Eftekhari, A.A.; Dafnomilis, G.; Lake, L.; Bruining, J. On the sustainability of CO_2 storage through CO_2–Enhanced oil recovery. *Appl. Energy* **2020**, *261*, 114467. [CrossRef]
2. Li, S.; Li, Z.; Dong, Q. Diffusion coefficients of supercritical CO_2 in oil-saturated cores under low permeability reservoir conditions. *J. CO_2 Util.* **2016**, *14*, 47–60. [CrossRef]
3. Zhang, L.; Wang, S.; Zhang, L.; Ren, S.; Guo, Q. Assessment of CO_2-EOR and its geo-storage potential in mature oil reservoirs, Shengli Oilfield, China. *Petrol. Explor. Dev.* **2009**, *36*, 737–742.
4. Dong, M.; Huang, S.; Srivastava, R. Effect of Solution Gas in Oil on CO_2 Minimum Miscibility Pressure. *Annu. Tech. Meet.* **1999**, *39*. [CrossRef]
5. Rathmell, J.; Stalkup, F.; Hassinger, R. A Laboratory Investigation of Miscible Displacement by Carbon Dioxide. *Soc. Pet. Eng. AIME Pap.* **1971**, *SPE3483*, 1–10. [CrossRef]
6. Zhang, K.; Jia, N.; Zeng, F.; Li, S.; Liu, L. A review of experimental methods for determining the Oil-Gas minimum miscibility pressures. *J. Pet. Sci. Eng.* **2019**, *183*. [CrossRef]
7. Glaso, O. Miscible Displacement: Recovery Tests with Nitrogen. *SPE Reserv. Eng.* **1990**, *5*, 61–68. [CrossRef]
8. Zhang, K.; Gu, Y. Two different technical criteria for determining the minimum miscibility pressures (MMPs) from the slim-tube and coreflood tests. *Fuel* **2015**, *161*, 146–156. [CrossRef]
9. Mogensen, K. A novel protocol for estimation of minimum miscibility pressure from slimtube experiments. *J. Pet. Sci. Eng.* **2016**, *146*, 545–551. [CrossRef]
10. Christiansen, R.L.; Haines, H.K. Rapid Measurement of Minimum Miscibility Pressure with the Rising-Bubble Apparatus. *SPE Reserv. Eng.* **1987**, *2*, 523–527. [CrossRef]
11. Zhang, K.; Jia, N.; Zeng, F. Application of predicted bubble-rising velocities for estimating the minimum miscibility pressures of the light crude oil–CO_2 systems with the rising bubble apparatus. *Fuel* **2018**, *220*, 412–419. [CrossRef]
12. Orr, F.M.; Jessen, K. An analysis of the vanishing interfacial tension technique for determination of minimum miscibility pressure. *Fluid Phase Equilibria* **2007**, *255*, 99–109. [CrossRef]
13. Ghorbani, M.; Momeni, A.; Safavi, S.; Gandomkar, A. Modified vanishing interfacial tension (VIT) test for CO_2–Oil minimum miscibility pressure (MMP) measurement. *J. Nat. Gas Sci. Eng.* **2014**, *20*, 92–98. [CrossRef]
14. Hemmati-Sarapardeh, A.; Ayatollahi, S.; Ghazanfari, M.-H.; Masihi, M. Experimental Determination of Interfacial Tension and Miscibility of the CO_2–Crude Oil System; Temperature, Pressure, and Composition Effects. *J. Chem. Eng. Data* **2014**, *59*, 61–69. [CrossRef]
15. Ayirala, S.C.; Rao, D.N. Application of the parachor model to the prediction of miscibility in multi-component hydrocarbon systems. *J. Phys. Condens. Matter* **2004**, *16*, S2177–S2186. [CrossRef]
16. Benham, A.; Dowden, W.; Kunzman, W. Miscible Fluid Displacement—Prediction of Miscibility. *Trans. AIME* **1960**, *219*, 229–237. [CrossRef]
17. Kuo, S. Prediction of Miscibility for the Enriched-Gas Drive Process. *SPE Annu. Tech. Conf. Exhib.* **1985**. [CrossRef]
18. Orr, F.J.; Silva, M. Effect of Oil Composition on Minimum Miscibility Pressure-Part 2: Correlation. *SPE Reserv. Eng.* **1987**, *2*, 479–491. [CrossRef]
19. Yuan, H.; Johns, R.T.; Egwuenu, A.M.; Dindoruk, B. Improved MMP Correlation for CO_2 Floods Using Analytical Theory. *SPE Reserv. Eval. Eng.* **2005**, *8*, 418–425. [CrossRef]
20. Valluri, M.K.; Mishra, S.; Schuetter, J. An improved correlation to estimate the minimum miscibility pressure of CO_2 in crude oils for carbon capture, utilization, and storage projects. *J. Pet. Sci. Eng.* **2017**, *158*, 408–415. [CrossRef]
21. Dong, M.; Huang, S.; Dyer, S.B.; Mourits, F.M. A comparison of CO_2 minimum miscibility pressure determinations for Weyburn crude oil. *J. Pet. Sci. Eng.* **2001**, *31*, 13–22. [CrossRef]
22. Chen, G.; Wang, X.; Liang, Z.; Gao, R.; Sema, T.; Luo, P.; Zeng, F.; Tontiwachwuthikul, P. Simulation of CO_2-Oil Minimum Miscibility Pressure (MMP) for CO_2 Enhanced Oil Recovery (EOR) using Neural Networks. *Energy Procedia* **2013**, *37*, 6877–6884. [CrossRef]
23. Chen, G.; Fu, K.; Liang, Z.; Sema, T.; Li, C.; Tontiwachwuthikul, P.; Idem, R. The genetic algorithm based back propagation neural network for MMP prediction in CO_2-EOR process. *Fuel* **2014**, *126*, 202–212. [CrossRef]
24. Tatar, A.; Shokrollahi, A.; Mesbah, M.; Rashid, S.; Arabloo, M.; Bahadori, A. Implementing Radial Basis Function Networks for modeling CO_2-reservoir oil minimum miscibility pressure. *J. Nat. Gas Sci. Eng.* **2013**, *15*, 82–92. [CrossRef]
25. Emera, M.K.; Sarma, H.K. Use of genetic algorithm to estimate CO_2–oil minimum miscibility pressure—A key parameter in design of CO_2 miscible flood. *J. Pet. Sci. Eng.* **2005**, *46*, 37–52. [CrossRef]
26. Sayyad, H.; Manshad, A.K.; Rostami, H. Application of hybrid neural particle swarm optimization algorithm for prediction of MMP. *Fuel* **2013**, *116*, 625–633. [CrossRef]
27. Chen, H.; Zhang, C.; Jia, N.; Duncan, I.; Yang, S.; Yang, Y. A machine learning model for predicting the minimum miscibility pressure of CO_2 and crude oil system based on a support vector machine algorithm approach. *Fuel* **2021**, *290*, 120048. [CrossRef]
28. Ghiasi, M.M.; Mohammadi, A.H.; Zendehboudi, S. Use of hybrid-ANFIS and ensemble methods to calculate minimum miscibility pressure of CO_2-reservoir oil system in miscible flooding process. *J. Mol. Liq.* **2021**, *331*, 115369. [CrossRef]
29. Li, D.; Li, X.; Zhang, Y.; Sun, L.; Yuan, S. Four Methods to Estimate Minimum Miscibility Pressure of CO_2-Oil Based on Machine Learning. *Chin. J. Chem.* **2019**, *37*, 1271–1278. [CrossRef]

30. Karkevandi-Talkhooncheh, A.; Hajirezaie, S.; Hemmati-Sarapardeh, A.; Husein, M.M.; Karan, K.; Sharifi, M. Application of adaptive neuro fuzzy interface system optimized with evolutionary algorithms for modeling CO_2-crude oil minimum miscibility pressure. *Fuel* **2017**, *205*, 34–45. [CrossRef]
31. Karkevandi-Talkhooncheh, A.; Rostami, A.; Sarapardeh, A.H.; Ahmadi, M.; Husein, M.M.; Dabir, B. Modeling minimum miscibility pressure during pure and impure CO_2 flooding using hybrid of radial basis function neural network and evolutionary techniques. *Fuel* **2018**, *220*, 270–282. [CrossRef]
32. Ekechukwu, G.K.; Falode, O.; Orodu, O.D. Improved Method for the Estimation of Minimum Miscibility Pressure for Pure and Impure CO_2–Crude Oil Systems Using Gaussian Process Machine Learning Approach. *J. Energy Resour. Technol.* **2020**, *142*, 1–14. [CrossRef]
33. Zendehboudi, S.; Rezaei, N.; Lohi, A. Applications of hybrid models in chemical, petroleum, and energy systems: A systematic review. *Appl. Energy* **2018**, *228*, 2539–2566. [CrossRef]
34. Zendehboudi, S.; Ahmadi, M.A.; Bahadori, A.; Shafiei, A.; Babadagli, T. A developed smart technique to predict minimum miscible pressure-eor implications. *Can. J. Chem. Eng.* **2013**, *91*, 1325–1337. [CrossRef]
35. Berendsen, H.J.C.; Van Der Spoel, D.; Van Drunen, R. GROMACS: A message-passing parallel molecular dynamics implementation. *Comput. Phys. Commun.* **1995**, *91*, 43–56. [CrossRef]
36. Hess, B.; Kutzner, C.; Van Der Spoel, D.; Lindahl, E. GROMACS 4: Algorithms for Highly Efficient, Load-Balanced, and Scalable Molecular Simulation. *J. Chem. Theory Comput.* **2008**, *4*, 435–447. [CrossRef]
37. Wang, J.; Wolf, R.M.; Caldwell, J.W.; Kollman, P.A.; Case, D.A. Case, "Development and testing of a general amber force field" Journal of Computational Chemistry(2004) 25(9) 1157-1174. *J. Comput. Chem.* **2004**, *26*, 114. [CrossRef]
38. Malde, A.; Zuo, L.; Breeze, M.; Stroet, M.; Poger, D.; Nair, P.; Oostenbrink, C.; Mark, A. An Automated Force Field Topology Builder (ATB) and Repository: Version 1.0. *J. Chem. Theory Comput.* **2011**, *7*, 4026–4037. [CrossRef] [PubMed]
39. Koziara, K.B.; Stroet, M.; Malde, A.; Mark, A.E. Testing and validation of the Automated Topology Builder (ATB) version 2.0: Prediction of hydration free enthalpies. *J. Comput. Mol. Des.* **2014**, *28*, 221–233. [CrossRef]
40. Bussi, G.; Donadio, D.; Parrinello, M. Canonical sampling through velocity rescaling. *J. Chem. Phys.* **2007**, *126*, 014101. [CrossRef]
41. Berendsen, H.J.C.; Postma, J.P.M.; Van Gunsteren, W.F.; DiNola, A.; Haak, J.R. Molecular dynamics with coupling to an external bath. *J. Chem. Phys.* **1984**, *81*, 3684–3690. [CrossRef]
42. Apostolakis, J.; Ferrara, P.; Caflisch, A. Calculation of conformational transitions and barriers in solvated systems: Application to the alanine dipeptide in water. *J. Chem. Phys.* **1999**, *110*, 2099–2108. [CrossRef]
43. Hess, B.; Bekker, H.; Berendsen, H.J.C.; Fraaije, J.G.E.M. LINCS: A linear constraint solver for molecular simulations. *J. Comput. Chem.* **1997**, *18*, 1463–1472. [CrossRef]
44. Darden, T.; York, D.; Pedersen, L. Particle mesh Ewald: An $N·\log(N)$ method for Ewald sums in large systems. *J. Chem. Phys.* **1993**, *98*, 10089–10092. [CrossRef]
45. Essmann, U.; Perera, L.; Berkowitz, M.; Darden, T.; Lee, H.; Pedersen, L.G. A smooth particle mesh Ewald method. *J. Chem. Phys.* **1995**, *103*, 8577–8593. [CrossRef]
46. Van der Spoel, D.; Lindahl, E.; Hess, B.; Van Buuren, A.; Apol, E.; Meulenhoff, P.; Tieleman, D.; Sijbers, A.; Feenstra, K.; van Drunen, R.; et al. Gromacs User Manual Version 4.0. *Manuals* **2005**. Available online: www.gromacs.org (accessed on 7 May 2021).
47. Teklebrhan, R.B.; Ge, L.; Bhattacharjee, S.; Xu, Z.; Sjöblom, J. Probing Structure–Nanoaggregation Relations of Polyaromatic Surfactants: A Molecular Dynamics Simulation and Dynamic Light Scattering Study. *J. Phys. Chem. B* **2012**, *116*, 5907–5918. [CrossRef]
48. De Lara, L.; Michelon, M.F.; Miranda, C.R. Molecular Dynamics Studies of Fluid/Oil Interfaces for Improved Oil Recovery Processes. *J. Phys. Chem. B* **2012**, *116*, 14667–14676. [CrossRef]
49. Kunieda, M.; Nakaoka, K.; Liang, Y.; Miranda, C.R.; Ueda, A.; Takahashi, S.; Okabe, H.; Matsuoka, T. Self-Accumulation of Aromatics at the Oil–Water Interface through Weak Hydrogen Bonding. *J. Am. Chem. Soc.* **2010**, *132*, 18281–18286. [CrossRef]
50. Song, S.; Zhang, H.; Sun, L.; Shi, J.; Cao, X.; Yuan, S. Molecular Dynamics Study on Aggregating Behavior of Asphaltene and Resin in Emulsified Heavy Oil Droplets with Sodium Dodecyl Sulfate. *Energy Fuels* **2018**, *32*, 12383–12393. [CrossRef]
51. Yu, H.; Lu, X.; Fu, W.; Wang, Y.; Xu, H.; Xie, Q.; Qu, X.; Lu, J. Determination of minimum near miscible pressure region during CO_2 and associated gas injection for tight oil reservoir in Ordos Basin, China. *Fuel* **2020**, *263*, 116737. [CrossRef]
52. Lee, J. *Effectiveness of Carbon Dioxide Displacement under Miscible and Immiscible Conditions*; Petroleum Recovery Inst.: Calgary, AB, Canada, 1979.
53. Alston, R.; Kokolis, G.; James, C. CO_2 Minimum Miscibility Pressure: A Correlation for Impure CO_2 Streams and Live Oil Systems. *Soc. Pet. Eng. J.* **1985**, *25*, 268–274. [CrossRef]
54. Shokir, E.M.E.-M. CO_2–oil minimum miscibility pressure model for impure and pure CO_2 streams. *J. Pet. Sci. Eng.* **2007**, *58*, 173–185. [CrossRef]
55. Cronquist, C. Carbon dioxide dynamic miscibility with light reservoir oils. In Proceedings of the Fourth Annual US DOE Symposium, Tulsa, OK, USA, 28 August 1978; pp. 28–30.
56. Glaso, O. Generalized Minimum Miscibility Pressure Correlation (includes associated papers 15845 and 16287). *Soc. Pet. Eng. J.* **1985**, *25*, 927–934. [CrossRef]
57. Yellig, W.F.; Metcalfe, R.S. Determination and Prediction of CO_2 Minimum Miscibility Pressures (includes associated paper 8876). *J. Pet. Technol.* **1980**, *32*, 160–168. [CrossRef]

58. Zolghadr, A.; Escrochi, M.; Ayatollahi, S. Temperature and Composition Effect on CO_2 Miscibility by Interfacial Tension Measurement. *J. Chem. Eng. Data* **2013**, *58*, 1168–1175. [CrossRef]
59. Zheng, L.; Ma, K.; Yuan, S.; Wang, F.; Dong, X.; Li, Y.; Du, D. Determination of the multiple-contact minimum miscibility pressure of CO_2/oil system using oil droplet volume measurement method. *J. Pet. Sci. Eng.* **2019**, 106578. [CrossRef]
60. Lashkarbolooki, M.; Eftekhari, M.J.; Najimi, S.; Ayatollahi, S. Minimum miscibility pressure of CO_2 and crude oil during CO_2 injection in the reservoir. *J. Supercrit. Fluids* **2017**, *127*, 121–128. [CrossRef]
61. Menouar, H. Discussion on Carbon Dioxide Minimum Miscibility Pressure Estimation: An Experimental Investigation. In Proceedings of the SPE Western Regional & AAPG Pacific Section Meeting 2013 Joint Technical Conference, Monterey, CA, USA, 19–25 April 2013.

Article

Demulsification of Heavy Oil-in-Water Emulsion by a Novel Janus Graphene Oxide Nanosheet: Experiments and Molecular Dynamic Simulations

Yingbiao Xu [1,2], Yefei Wang [1,*], Tingyi Wang [2], Lingyu Zhang [2], Mingming Xu [2] and Han Jia [1,*]

[1] Key Laboratory of Unconventional Oil & Gas Development, China University of Petroleum (East China), Ministry of Education, Qingdao 266580, China; xuyingbiao.slyt@sinopec.com
[2] Technology Inspection Center, Shengli Oilfield Company, SINOPEC, Dongying 257000, China; wangtingyi180.slyt@sinopec.com (T.W.); zhangly639.slyt@sinopec.com (L.Z.); xumingming.slyt@sinopec.com (M.X.)
* Correspondence: wangyf@upc.edu.cn (Y.W.); jiahan@upc.edu.cn (H.J.)

Abstract: Various nanoparticles have been applied as chemical demulsifiers to separate the crude-oil-in-water emulsion in the petroleum industry, including graphene oxide (GO). In this study, the Janus amphiphilic graphene oxide (JGO) was prepared by asymmetrical chemical modification on one side of the GO surface with n-octylamine. The JGO structure was verified by Fourier-transform infrared spectra (FTIR), transmission electron microscopy (TEM), and contact angle measurements. Compared with GO, JGO showed a superior ability to break the heavy oil-in-water emulsion with a demulsification efficiency reaching up to 98.25% at the optimal concentration (40 mg/L). The effects of pH and temperature on the JGO's demulsification efficiency were also investigated. Based on the results of interfacial dilatational rheology measurement and molecular dynamic simulation, it was speculated that the intensive interaction between JGO and asphaltenes should be responsible for the excellent demulsification performance of JGO. This work not only provided a potential high-performance demulsifier for the separation of crude-oil-in-water emulsion, but also proposed novel insights to the mechanism of GO-based demulsifiers.

Keywords: heavy oil-in-water emulsion; demulsification; Janus graphene oxide; molecular dynamic simulation

1. Introduction

The treatment of liquids produced from crude oil production has become a great challenge in the petroleum industry [1–7]. With excessive exploitation, most oilfields have reached the high water cut stage, a circumstance in which crude-oil-in-water emulsion was usually generated in the produced liquids [8,9]. In the case of heavy oil production, the ubiquitous asphaltene acts as natural emulsifier to readily adsorb on the oil/water interface and dramatically enhance the emulsion stability [10–12]. The extremely stable emulsion can cause serious problems to the downstream process, such as the generation of large amounts of polluting oily wastewater [13,14]. Therefore, it is an urgent issue to develop a highly efficient demulsifier to separate the oil from the emulsion generated in heavy oil exploitation.

The proposed strategies for oil–water separation mainly include adsorption [15], coalescence technology [16], membrane separation [17], gravity separation [18], flotation [19], and ultra-centrifugation [20]. Among these methods, chemical demulsification is widely applied owing to the tunable structure of demulsifier, high efficiency, and convenient operation [21]. Various chemical demulsifiers have been developed in last few decades, such as block copolymer [22], polysiloxane [23], hyperbranched polymer [24], dendrimer [25], and ionic liquids [26]. Chemical demulsifiers are widely used to break water-in-oil emulsions

during the early stage of oilfield development. With the continuous development of oilfields, the study of oil-in-water emulsions has attracted a great deal of attention. While the increasing concern on environmental protection restricts the practical application of these chemicals [27], the research and development of novel demulsifiers is still a challenging and study-worthy topic for petroleum engineers.

Recently, nanomaterials have been regarded as another crucial category of chemical demulsifiers [28]. Nikkhah et al., verified that the settling time of nano-titania was significantly less than conventional chemical demulsifiers [29]. Zhang et al., fabricated the cyclodextrin-modified magnetic composite particles (M-CD) and evaluated its demulsification performance. M-CD could effectively separate various types of emulsions and exhibit excellent recycling ability [30]. Ren's group systematically studied the demulsification performance of graphene oxide (GO) and its derivatives [31–33]. They concluded that the asphaltenes could effectively stabilize the oil-in-water emulsions under certain conditions, which created a large number of problems in practical production of the oilfield, and they proposed that the amphiphilic GO-based materials could adsorb on the oil/water interface to promote the coalescence of emulsion droplets.

It is well accepted that the amphiphilicity of GO is relatively weak due to its high content of hydrophilic oxygen-containing groups [34], while these active groups also provide the possibility for chemically hydrophobic modification of GO to improve its interfacial activity [35]. Janus graphene oxide (JGO) refers to GO with one side modified and has been applied in various fields due to its unique structure and properties [36,37]. For the petroleum industry, Luo et al., reported the application of JGO as a promising nano-fluid for enhanced oil recovery [38]. JGO with improved amphiphilicity can easily migrate to oil/water interface and interact with asphaltene, which may exhibit a better demulsification performance than GO.

Today, molecular dynamics (MD) simulations are employed extensively in the petroleum industry, which can reveal some interfacial properties of multiphase systems at the molecular level [39–46]. Stephan et al., investigated the vapor–liquid interface properties of binary mixtures (cyclohexane + CO_2) via MD simulations [47]. Chakraborty et al., used MD simulations to study the vapor–liquid interface properties of n-heptane/nitrogen at different temperatures and pressures [48]. Lian et al., discussed the interaction of zwitterionic surfactant with various components of crude oil (asphaltene, resin, saturate, and aromatic) at the molecular level [49]. Liu et al., researched the emulsification and demulsification capabilities of a gas switchable surfactant through molecular dynamics simulations [50].

In this study, the JGO was synthesized by grafting n-octylamine on one surface of GO, and the successful modification was verified by FTIR, TEM, and contact angle measurements. Then, the demulsification efficiency of JGO was systematically investigated and compared with GO, including the effects of dosage, pH, and temperature. Based on the effects of JGO on the interfacial rheological properties and molecular dynamic (MD) simulation on the interaction between JGO and asphaltene, the passible demulsification mechanism of JGO was proposed.

2. Results and Discussion

2.1. Characterization of JGO

The FTIR spectra of GO and JGO were measured to verify the conjunction of n-octylamine on the JGO surface (Figure 1). The additional peaks between 3100 and 2800 cm^{-1} for the JGO sample were assigned to the stretching vibration of C-H (-CH_2- and -CH_3), which directly indicates the existence of n-octylamine with the hydrocarbon chain. Moreover, the weaker peaks at 1730 cm^{-1} (C=O) and 1208 cm^{-1} (C-O-C) in JGO's curve than those in GO's curve reflect that the n-octylamine was successfully grafted on the JGO surface via the reaction between amido group and epoxy/carboxyl groups.

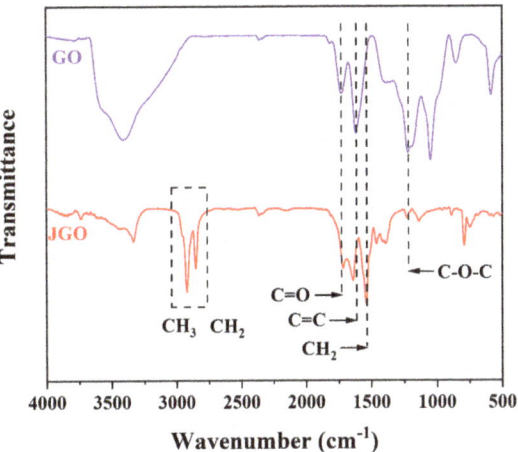

Figure 1. FTIR spectra of GO and JGO.

Figure 2 presents the TEM images of GO and JGO. It is obvious that the modified n-octylamine hardly changed the morphology and size of GO-based nanosheets, but only resulted in more overlapped parts in JGO due to the additional hydrophobic interaction. To further demonstrate the Janus structure of JGO, the JGO interfacial film constructed at the octane/water interface was transferred to the glass substrate to measure water contact angles of both sides of the JGO. As shown in Figure S1, the unmodified side of JGO was relatively hydrophilic with a water contact angle of 32°, which was attributed to the abundant oxygen-containing groups. For the n-octylamine grafted side of JGO, the water contact angle increased to 85° due to the hydrophobic hydrocarbon chain. The opposite affinity to water for both sides of JGO directly confirmed the generation of the Janus structure.

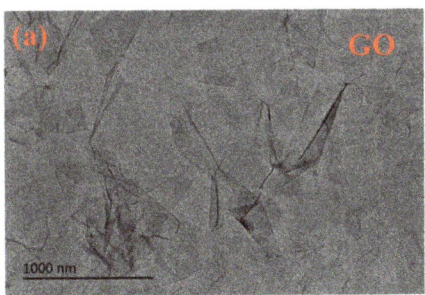
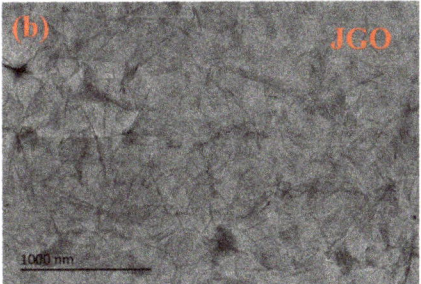

Figure 2. TEM images of GO (**a**) and JGO (**b**).

2.2. Demulsification Efficiency Tests

The demulsification efficiency of GO and JGO for the stable heavy oil-in-water emulsions (prepared with 5% heavy oil) was evaluated by the bottle test (Figure 3). The blank sample exhibited excellent stability, with almost no separated oil from the emulsion (Figure S2). The optimal demulsification efficiency of JGO was achieved at the much lower concentration (40 mg/L), and the oil content was sharply reduced to 825 mg/L. The emulsion color became lighter with additional demulsifiers (GO or JGO), and the emulsion droplets spontaneously aggregated to form a separated oil layer. Meanwhile, there were flocculent aggregates on the top of the water phase. The higher demulsification efficiency of JGO (98.25%) than that of GO (92.5%) indicates that the asymmetrically modified n-octylamine can greatly improve the demulsification performance of GO-based nanosheets. Higher concentrations of JGO slightly decreased the demulsification efficiency, which may be ascribed to the adsorbed oil on the JGO surface.

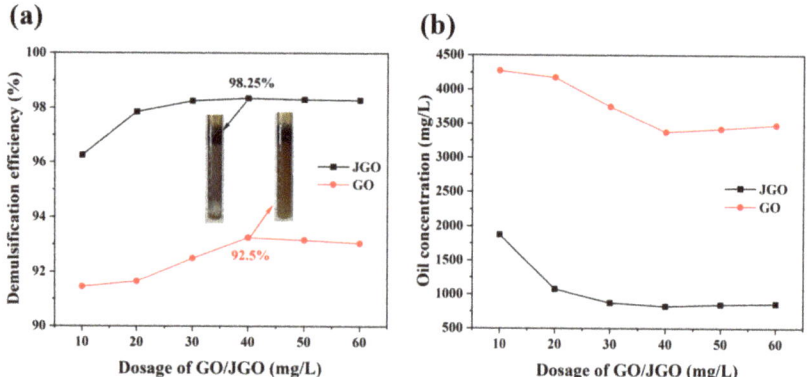

Figure 3. The demulsification efficiency (**a**) and oil concentrations (**b**) of GO and JGO with different dosages.

To systematically investigate the demulsification performances of JGO (30 mg/L) under different conditions, the effects of pH and temperature were studied via a series of experiments (Figure 4). It is well known that the solution pH value can significantly affect the surface physiochemical properties of the demulsifiers to impact their performances [27]. For JGO, the demulsification efficiency at acidic and natural conditions was excellent, but the alkaline environment was adverse to its performance (Figure 4a). On the one hand, the organic acid in crude oil would react with alkali to generate a surface-active substance, which could largely improve the emulsion stability. On the other hand, the variation of zeta potential of JGO and emulsions at different pH values should be the other factor for the pH-dependent demulsification efficiency of JGO (Figure S3). With increasing pH values, both JGO and emulsion droplets became more negatively charged. Therefore, the intensive electrostatic repulsion between JGO and emulsion droplets at the alkaline condition (pH $\geq$ 8) would weaken the demulsification efficiency of JGO. In addition, the demulsification efficiency of JGO slightly increased with increased temperature (Figure 4b). Although the higher temperature may disturb the interaction between JGO and emulsion droplets, the accelerated thermal movements of emulsion droplets facilitated their coalescence. More importantly, it was verified that JGO can effectively separate heavy oil-in-water emulsions in a wide temperature range.

2.3. Demulsification Mechanism of JGO

For the emulsions prepared in this study, the emulsion stability is mainly dependent on the surface-active substance in crude oil, asphaltene in particular. Asphaltene could fabricate a protective film at the oil/water interface to prevent the coalescence of the emulsion droplets [31]. Therefore, the additional demulsifier should primarily destroy the asphaltene film to achieve demulsification. To investigate the effects of GO and JGO on the interfacial properties of the crude oil/water interface, the interfacial dilatational rheology experiment was conducted. As shown in Figure 5, the dilatational modulus of crude oil/water system was 12.32 mN/m, which should be attributed to the interfacial adsorption of asphaltene. The addition of GO or JGO in the water phase causes the evident increase of the dilatational modulus in different degrees. GO with the hydrophilic edge and the hydrophobic plane is generally regarded as the amphiphile [51], whereas the amphiphilicity of JGO was further improved by the asymmetric modification with n-octylamine. Then, the interfacial film is the combination of asphaltene and GO or JGO.

For the blank and GO system, the interfacial film was basically elastic, with the phase angle around 10°, while the phase angle of JGO system increased dramatically, indicating the typical viscoelastic property of the interfacial film. The smaller E and larger E'' in the JGO system could be ascribed to the Janus structure. Considering GO and JGO nanosheets are 2D materials with an average literal size around 500 nm, the adsorbed GO or JGO at the interface should be overlapped with each other. The overlapped parts of JGO nanosheets may not be as rigid as those of GO nanosheets due to the hinderance of n-octylamine. Therefore, the JGO adsorbed interface of the emulsion would be much easier to deform, leading to the lower E and more distinct viscous characteristic.

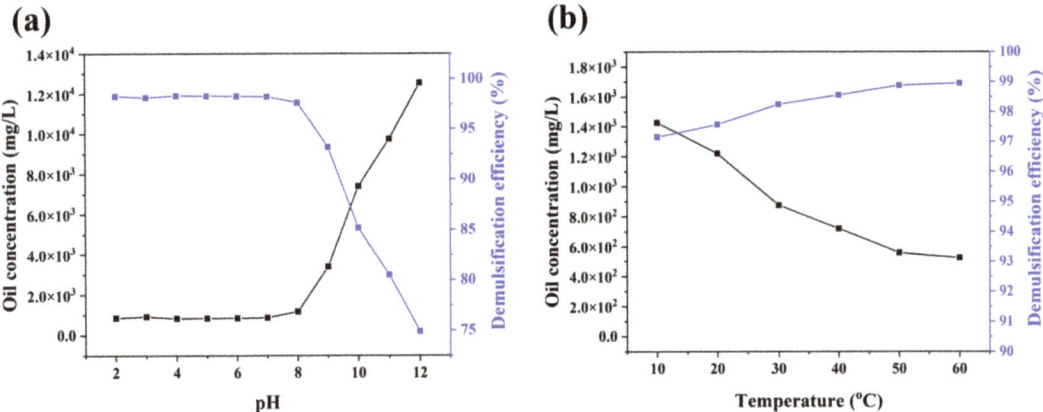

Figure 4. The effects of pH (**a**) and temperature (**b**) on the demulsification performance of JGO (30 mg/L).

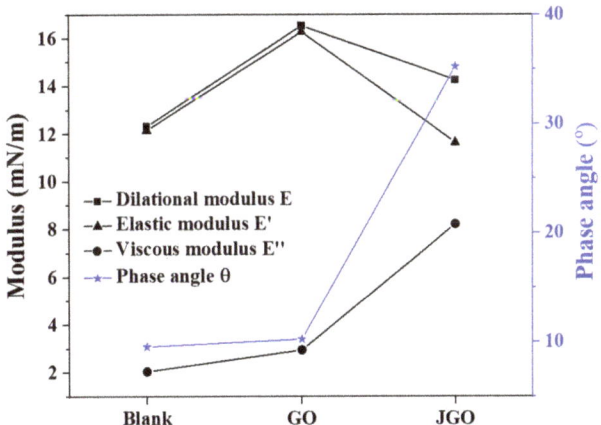

Figure 5. Effects of GO and JGO on the interfacial rheology of crude oil/water interface.

Based on the results of interfacial rheology, the possible demulsification mechanism of JGO could be proposed (Figure 6). The adsorbed asphaltene molecules constructed the protective film at the oil/water interface, generating a new phase to stabilize the heavy oil-in-water emulsion from the thermodynamic aspect (Figure 6a) [41,52]. After JGO was added to the emulsion system, it could disperse well in the water phase (Figure 6b). Then, JGO with improved amphiphilicity could adsorb at the oil/water interface to interact with asphaltene (Figure 6c). During the shaking, the emulsion droplets collided with each other (Figure 6d). The intensive interaction between the amphiphilic JGO and the surface-active substance asphaltene could destroy the original interfacial film, and parts of asphaltene desorbed from the oil/water interface, leading to the coalescence of smaller ones without the protective film (Figure 6e). Finally, after settling, the larger oil droplets moved upward to form a separated oil layer, and the JGO/asphaltene aggregation also moved to the upper region of the water phase due to the highly improved hydrophobicity (Figures 3a and 6f). The greater amphiphilicity of JGO should be responsible for its higher demulsification efficiency than GO.

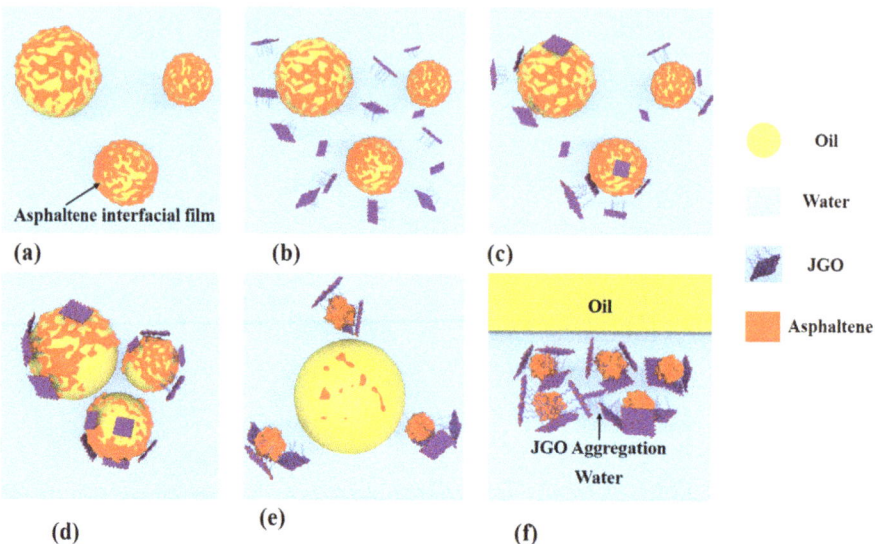

Figure 6. Schematic of the demulsification mechanism of JGO; (**a**) stable emulsion; (**b**) adding JGO; (**c**) adsorption of JGO; (**d**) shaking; (**e**) aggregation of oil droplets; (**f**) after settling.

2.4. Verification of the Mechanism via Molecular Dynamic Simulation

To verify the mechanism of the JGO as a high-performance demulsifier, the behavior of GO and JGO in the crude oil/water system was simulated via molecular dynamic (MD) simulation. As shown in Figure S4, GO and JGO nanosheets were randomly placed in the water phase in the initial configuration, which could achieve the dynamic equilibrium after 50 ns MD simulation (Figure S5). The Gibbs partition surfaces of two systems were located at ~5.1 nm and ~14.0 nm, and the density peaks of asphaltene (including asphaltene-1, asphaltene-2, and asphaltene-3 in Scheme S1), GO, and JGO in two systems were all around the Gibbs partition surface, indicating that the GO-based nanosheets can spontaneously adsorb at the crude oil/water interface and interact with asphaltene (Figure 7). Compared with GO, there was more JGO adsorbed at the interface with the n-octylamine modified side toward the oil phase. Meanwhile, the density peak of asphaltene in the JGO system was much greater than that in the GO system.

To quantificationally investigate the interaction energy between GO/JGO and asphaltene, the nonbonded interaction, including Lennard–Jones potential and Coulomb potential, during the whole simulation process was extracted (Figure 8a,b). The much larger total interaction energy between JGO and asphaltene (~−650 kJ/mol) than that between GO and asphaltene (~−250 kJ/mol) further confirmed their more intensive interaction (Figure 8c). It is worth noting that the much larger interaction energy in the JGO-asphaltene system was mainly derived from the stronger Lennard–Jones potential, which should be ascribed to the hydrophobic interaction among the alkyl chains on both JGO and asphaltene.

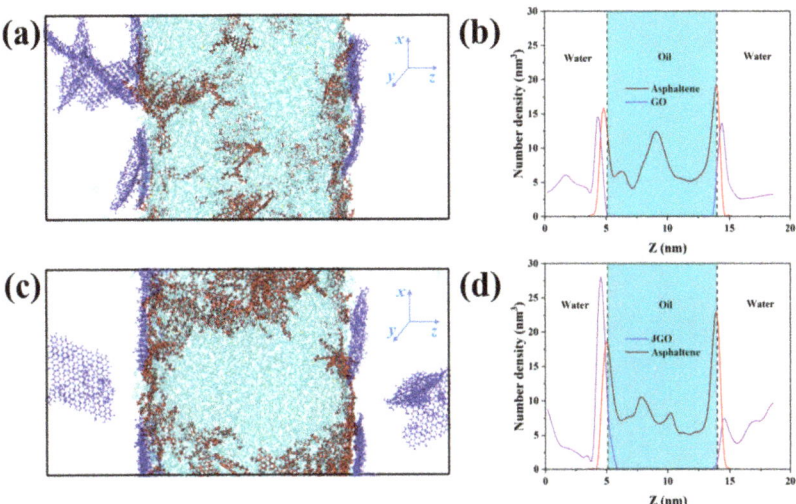

Figure 7. Final simulation configuration (at 50 ns) of GO (**a**) and JGO (**c**) in the oil/water system. Oil components are colored in cyan except asphaltene which is colored in red, and GO and JGO are colored in blue. Water molecules are not shown for clarity. Number density of three asphaltene in oil, GO (**b**), and JGO (**d**) along the z-direction. Gray dash lines in (**b**,**d**) represent the Gibbs partition surface.

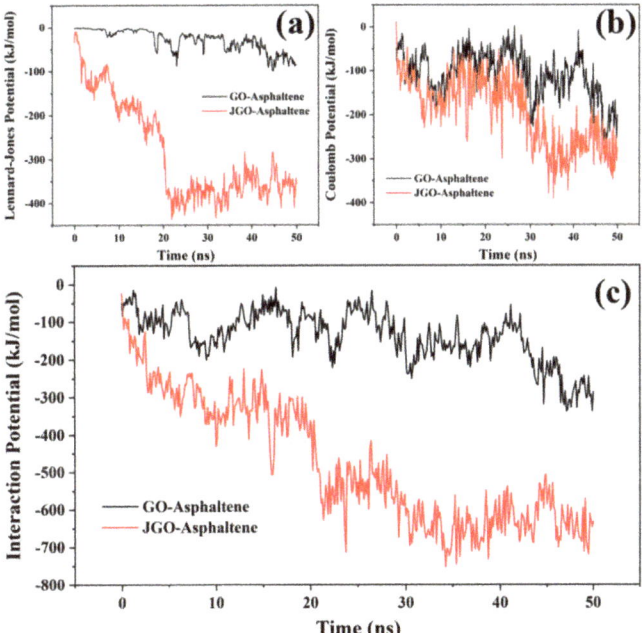

Figure 8. Intermolecular interaction energy between GO/JGO and asphaltene at the oil/water interface: Lennard–Jones potential (**a**), Coulomb potential (**b**), and total interaction energy (**c**).

As mentioned in the proposed mechanism, the asphaltene-constructed protective film was essential for the emulsion stability. Then, the distribution characteristic of asphaltene at the crude oil/water interface was further analyzed by calculating and plotting the two-dimensional (2D) number density map in the XY plane (Figure 9). The color distribution of the 2D number density maps directly indicates the asphaltene distribution and compactness in the GO-free, GO, and JGO systems. In the GO-free system, the more green and less yellow/red areas reflect the uniform distribution of asphaltene at the crude oil/water interface (Figure 9a). Compared with that in the GO system, the generation of dark red and red areas represents the much greater asphaltene density at the local interface (Figure 9b,c). On the one hand, the uneven distribution of asphaltene should be attributed to the constraint of the intensive interaction between JGO and asphaltene. On the other hand, the uneven distribution of asphaltene may cause the evident weakness in the protective film, resulting in the easy coalescence of the emulsion droplets.

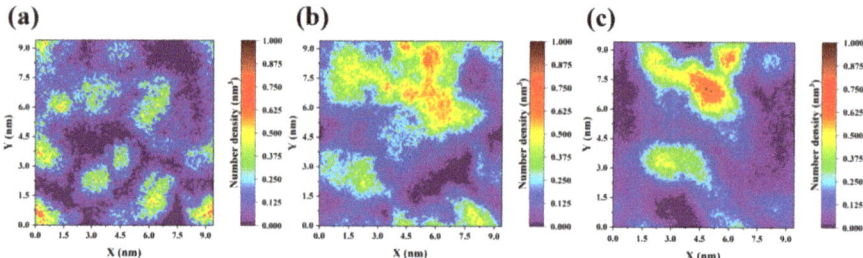

Figure 9. Two-dimensional number density maps of asphaltene distribution in the XY plane at the oil/water interface in the GO-free (**a**), GO (**b**), and JGO (**c**) systems.

3. Experimental

3.1. Materials

Graphene oxide (GO) was purchased from Shengzhen Turing Evolution Technical Company, China. Paraffin wax (with a melting point around 58−60 °C), kerosene, n-octylamine (>98%), ethanol (>99.5%), n-octane (>99%), and NaCl (>99.5%) were products from Shanghai Aladdin Biochemical Technology Co., Ltd., Shanghai, China.

3.2. Preparation of JGO

The method applied for the preparation of JGO was reported by Luo et al. [53]. First, 200 mL GO aqueous solution (1 mg/mL), 6 g NaCl, and 50 g paraffin wax were heated to 75 °C and stirred with a homogenizer at 10,000 rpm for 10 min to form emulsions, which were cooled to ambient temperature and filtered to obtain GO-coated wax particles. After the GO-coated wax particles were washed by NaOH solution (pH ~10), deionized (DI) water, and ethanol successively, the wax particles were added into the n-octylamine ethanol solution (0.4489 mg/mL) and magnetically stirred overnight. Then, the wax particles were washed with ethanol and dissolved in toluene to remove the wax, and JGO was obtained by centrifugation. Finally, JGO was dried at 60 °C and then dispersed in DI water at certain concentrations.

3.3. Characterization

The Fourier-transform infrared spectra (FTIR) of GO and JGO were recorded by a PerkinElmer Spectrum Two spectrometer (PerkinElmer, Waltham, MA, USA). Transmission electron microscopy (TEM) images of GO and JGO were obtained with a JEOL JEM-1400 transmission electron microscope (JEOL, Tokyo, Japan).

The contact angle measurements were conducted to the affinity of both sides of JGO. First, the interfacial film was fabricated by shaking the glass tube filled with JGO aqueous solution and n-octane, then the n-octane was removed by evaporation. To measure the contact angle of the n-octylamine modified (hydrophobic) side of JGO, the pre-cleaned glass slide was lifted below the JGO interfacial film to deposit the film on the glass substrate.

To measure the contact angle of the unmodified (hydrophilic) side of JGO, the pre-cleaned glass substrate was pressed onto the interfacial film and rotated immediately. After the obtained glass substrates were dried at 40 °C, the water contact angle measurement was conducted with a contact angle goniometer (JC2000D5M, Zhongchen, China).

3.4. Demulsification Efficiency Test

The heavy oil used in this study was obtained from Shengli oil field. To prepare stable heavy oil-in-water emulsions, 5 g heavy oil and 95 g NaCl aqueous solution (3000 mg/L) were mixed at 60 °C and stirred with a FJ200-S homogenizer at 12,000 rpm for 10 min. For the demulsification test, a certain amount of JGO (or GO as control sample) aqueous solution was added into the prepared emulsion stored in the measuring cylinder. Then, the cylinder was shaken for 1 min to ensure the thorough mixture of JGO and the emulsion. Then, the cylinder was placed at ambient temperature and the oil/water separation process was recorded by a camera. The oil content in the water phase after demulsification process was determined by UV–Vis spectrophotometer. The standard curve of oil content as a function of absorbance at 256 nm was obtained by measuring the absorbance of a series of mixture with different oil content. The demulsification efficiency (E_D) was calculated by the following equation:

$$E_D = \frac{C_0 - C_1}{C_0} \times 100\% \tag{1}$$

where C_0 is the initial oil content before demulsification, and C_1 is the oil content after demulsification.

3.5. Interfacial Dilatational Rheology Measurement

The interfacial property was measured by a dynamic interfacial oscillatory drop tensionmeter (Tracker, Teclis, France). A drop of crude oil (20 μL) was injected into the solutions through an inverted needle, and the volume of the oil drop changed in sinusoidal oscillatory motion, which was achieved by a motor system connected to the syringe. Based on the shape variation of the oil drop recorded by a CDD digital camera, the IFT was calculated according to the Laplace–Young equation and the Plane hydrostatic equation by computer software. At the same time, the dilatational modulus (E) was calculated by the software using the following equation [54]:

$$E = \frac{d\gamma}{d\ln A} \tag{2}$$

where γ is the interfacial tension (mN/m), and A is the interface area (m^2).

The dilatational modulus can also be expressed in plural form:

$$E = E' + E'' = E' + i\omega\eta_d \tag{3}$$

where E' is the elastic modulus, E'' is the viscous modulus, ω is the interfacial dilatation frequency, and η_d is the interfacial dilatation viscosity.

3.6. Simulation Method
3.6.1. Construction of Crude Oil/Water System

The crude oil model in this simulation was constructed based on the typical heavy oil model, which is composed by asphaltene, resin, aromatic, and saturate components [55,56]. Based on previous literature, three typical asphaltenes molecules and six types of resin molecules were applied in this crude oil model [55,56]. The aromatic components mainly include toluene and benzene. As for saturates, cycloheptane, cyclohexane, nonane, octane, heptane, and hexane were incorporated. The detailed information of the crude oil components was listed in Table S1, and the molecular structures of asphaltene and resin were depicted in Scheme S1.

To obtain the crude oil model, all component molecules were added randomly into a cubic box (12 × 12 × 12 nm^3), and 20 ns MD simulation under NPT ensembles was performed to achieve the reasonable density. The final crude oil model was obtained with the cubic box size of 9.4 × 9.4 × 9.4 nm^3, and the box was extended along z-direction to 9.4 × 9.4 × 18.8 nm^3. The empty parts of the extended crude oil box were filled with water molecules and counterions (Na$^+$) to establish the initial oil/water interface. Finally, 50 ns MD simulation under NPT ensemble was carried out to achieve the dynamic equilibrium of crude oil/water simulation system.

3.6.2. Construction of GO and JGO in Crude Oil/Water System

The single layer of GO was built with 286 carbon atoms as the substrate, 6 carboxyl groups on the edge, 25 epoxy groups on each side, 10 hydroxyl groups on each side, and 12 hydroxyl groups on the edge [57]. Based on the model of GO, six n-octylamine molecules were attached on one surface of GO via the ring-opening of epoxy groups to form JGO. The molecular structures of GO and JGO were illustrated in Scheme S2. Then, 10 GO/JGO were randomly inserted into the water phase of a pre-equilibrated crude oil/water system. The 50 ns MD simulation under NPT ensemble was carried out to achieve the dynamics equilibrium of the simulation system.

3.6.3. MD Simulation Methods

The GROMACS (version 2019.6) software package was used to perform MD simulations with optimized potentials for liquids simulation-all atom (OPLS-AA) force field [58]. Molecular parameter sets were generated from the LigParGen web server [59]. The TIP4P model was adopted to describe the water molecule. The steepest descent method was used to minimize the simulation system, and the convergence criterion of energy minimization was 50 kJ/(mol·nm). Then, the MD simulation with NPT ensemble at 1 atm and 303 K was carried out for each system. In the simulation, the velocity rescaling thermostat and Berendsen (first 10 ns) + Parrinello-Rahman (last 40 ns) pressure coupling were employed as temperature coupling method and pressure coupling method, respectively [60–62]. The LINCS algorithm was used to constrain the bonds with H atoms [63]. The cut off scheme was applied in the van der Waals (vdW) interaction. Coulomb interaction was computed using the Particle-Mesh Ewald (PME) method [64]. The simulation time step was 2 fs and trajectories were saved every 10 ps for further analysis, which was visualized by visual molecular dynamics (VMD) software [65].

4. Conclusions

In conclusion, we successfully synthesized n-octylamine-modified JGO. Compared with GO, JGO exhibited superior ability to effectively separate the heavy oil-in-water emulsion with a demulsification efficiency as high as 98.25% at the much lower concentration (40 mg/L). The interfacial dilatational rheology measurements demonstrated that the additional JGO to heavy oil-in-water emulsion systems could improve the viscoelasticity of the elastic oil/water interface, which facilitated the deformation of the interface and the destruction of the protective film. Moreover, the MD simulation further verified the more intensive adsorption of JGO on the crude oil/water interface and the stronger interaction between JGO and asphaltene in comparison to the GO system. Therefore, it is believed that the remarkable demulsification ability of JGO should be attributed to its powerful attraction with asphaltene, leading to the easily deformable oil/water interface and the uneven distribution of asphaltene. This study indicates that JGO could be applied as a high-performance demulsifier to separate heavy oil-in-water emulsions in the oil industry, and the proposed mechanism of JGO could be inspiring for a new strategy of chemical demulsifier design.

Supplementary Materials: The following supporting information can be downloaded at: https://www.mdpi.com/article/10.3390/molecules27072191/s1, Scheme S1: Molecular structures of different asphaltenes and resins; Scheme S2: Molecular structures of GO and JGO; Figure S1: The water contact angle of the unmodified side of JGO (a) and n-octylamine grafted side of JGO (b); Figure S2: Demulsification performance of GO and JGO; Figure S3: Zeta potential of JGO and oil droplets with increasing pH values; Figure S4: The initial configuration of GO and JGO in the crude oil/water system, oil components are colored in cyan excepting asphaltenes which are colored in red, and GO and JGO are colored in blue. Water molecules are not shown for clarity; Figure S5: Total energy curves in the simulation process with 10 GO (a) and JGO (b) randomly inserted into oil/water systems; Table S1: Compositions of the crude oil model.

Author Contributions: Data curation, T.W.; Formal analysis, M.X.; Funding acquisition, H.J.; Investigation, Y.X.; Methodology, Y.W. and L.Z.; Software, Y.W.; Supervision, H.J.; Writing–original draft, Y.X. All authors have read and agreed to the published version of the manuscript.

Funding: The authors are grateful for funding from the National Natural Science Foundation of China (Grant No. 52174053).

Institutional Review Board Statement: Not applicable.

Informed Consent Statement: Not applicable.

Data Availability Statement: The data presented in this study are available on request from the corresponding authors.

Conflicts of Interest: The authors declare no conflict of interest.

References

1. Atta, A.M.; Al-Lohedan, H.A.; Abdullah, M.M.S. Dipoles poly(ionic liquids) based on 2-acrylamido-2-methylpropane sulfonic acid-co-hydroxyethyl methacrylate for demulsification of crude oil water emulsions. *J. Mol. Liq.* **2016**, *222*, 680–690. [CrossRef]
2. Chen, Z.; Peng, J.; Ge, L.; Xu, Z. Demulsifying water-in-oil emulsions by ethyl cellulose demulsifiers studied using focused beam reflectance measurement. *Chem. Eng. Sci.* **2015**, *130*, 254–263. [CrossRef]
3. Kang, W.; Guo, L.; Fan, H.; Meng, L.; Li, Y. Flocculation, coalescence and migration of dispersed phase droplets and oil-water separation in heavy oil emulsion. *J. Pet. Sci. Eng.* **2012**, *81*, 177–181. [CrossRef]
4. Lv, X.; Song, Z.; Yu, J.; Su, Y.; Zhao, X.; Sun, J.; Mao, Y.; Wang, W. Study on the demulsification of refinery oily sludge enhanced by microwave irradiation. *Fuel* **2020**, *279*, 118417. [CrossRef]
5. Wang, Z.; Gu, S.; Zhou, L. Research on the static experiment of super heavy crude oil demulsification and dehydration using ultrasonic wave and audible sound wave at high temperatures. *Ultrason. Sonochem.* **2018**, *40*, 1014–1020. [CrossRef] [PubMed]
6. Yan, C.; Han, J.; Huang, C.; Mu, T. Demulsification of Water-in-Oil Emulsions for the Petroleum Industry by using Alternating Copolymers. *Energy Technol.* **2014**, *2*, 618–624. [CrossRef]
7. Zhao, F.; Tian, Z.; Yu, Z.; Shang, H.; Wu, Y.; Zhang, Y. Research status and analysis of stabilization mechanisms and demulsification methods of heavy oil emulsions. *Energy Sci. Eng.* **2020**, *8*, 4158–4177.
8. Liu, R.; Lu, Y.; Pu, W.; Lian, K.; Sun, L.; Du, D.; Song, Y.; Sheng, J.J. Low-Energy Emulsification of Oil-in-Water Emulsions with Self-Regulating Mobility via a Nanoparticle Surfactant. *Ind. Eng. Chem. Res.* **2020**, *59*, 18396–18411. [CrossRef]
9. Lobo, A.; Cambiella, A.; Benito, J.M.; Pazos, C.; Coca, J. Ultrafiltration of oil-in-water emulsions with ceramic membranes: Influence of pH and crossflow velocity. *J. Membr. Sci.* **2006**, *278*, 328–334. [CrossRef]
10. Lv, G.; Gao, F.; Liu, G.; Yuan, S. The properties of asphaltene at the oil-water interface: A molecular dynamics simulation. *Colloid Surf. A Physicochem. Eng. Asp.* **2017**, *515*, 34–40. [CrossRef]
11. Shi, C.; Zhang, L.; Xie, L.; Lu, X.; Liu, Q.; He, J.; Mantilla, C.A.; Van den Berg, F.G.A.; Zeng, H. Surface Interaction of Water-in-Oil Emulsion Droplets with Interfacially Active Asphaltenes. *Langmuir* **2017**, *33*, 1265–1274. [CrossRef] [PubMed]
12. Simon, S.; Sjöblom, J.; Wei, D. Interfacial and Emulsion Stabilizing Properties of Indigenous Acidic and Esterified Asphaltenes. *J. Dispers. Sci. Technol.* **2016**, *37*, 1751–1759. [CrossRef]
13. Dai, Q.; Chung, K.H. Hot water extraction process mechanism using model oil sands. *Fuel* **1996**, *75*, 220–226. [CrossRef]
14. Masliyah, J.; Zhou, Z.J.; Xu, Z.H.; Czarnecki, J.; Hamza, H. Understanding water-based bitumen extraction from athabasca oil sands. *Can. J. Chem. Eng.* **2004**, *82*, 628–654. [CrossRef]
15. Sokker, H.H.; El-Sawy, N.M.; Hassan, M.A.; El-Anadouli, B.E. Adsorption of crude oil from aqueous solution by hydrogel of chitosan based polyacrylamide prepared by radiation induced graft polymerization. *J. Hazard. Mater.* **2011**, *190*, 359–365. [CrossRef]
16. Srinivasan, A.; Viraraghavan, T.; Ng, K.T.W. Coalescence/Filtration of an Oil-In-Water Emulsion in an Immobilized Mucor rouxii Biomass Bed. *Sep. Sci. Technol.* **2012**, *47*, 2241–2249.

17. Li, H.; Mu, P.; Li, J.; Wang, Q. Inverse desert beetle-like ZIF-8/PAN composite nanofibrous membrane for highly efficient separation of oil-in-water emulsions. *J. Mater. Chem. A* **2021**, *9*, 4167–4175. [CrossRef]
18. Frising, T.; Noïk, C.; Dalmazzone, C. The liquid/liquid sedimentation process: From droplet coalescence to technologically enhanced water/oil emulsion gravity separators: A review. *J. Dispers. Sci. Technol.* **2006**, *27*, 1035–1057. [CrossRef]
19. Moosai, R.; Dawe, R.A. Gas attachment of oil droplets for gas flotation for oily wastewater cleanup. *Sep. Purif. Technol.* **2003**, *33*, 303–314. [CrossRef]
20. Fan, C.; Ma, R.; Wang, Y.; Luo, J. Demulsification of Oil-in-Water Emulsions in a Novel Rotating Microchannel. *Ind. Eng. Chem. Res.* **2020**, *59*, 8335–8345. [CrossRef]
21. Dhandhi, Y.; Chaudhari, R.K.; Naiya, T.K. Development in separation of oilfield emulsion toward green technology—A comprehensive review. *Sep. Sci. Technol.* **2021**, 1–27. [CrossRef]
22. Atta, A.M.; Fadda, A.A.; Abdel-Rahman, A.A.H.; Ismail, H.S.; Fouad, R.R. Application of New Modified Poly(ethylene Oxide)-Block-Poly(propylene oxide)-Block-Poly(ethylene oxide) Copolymers as Demulsifier for Petroleum Crude Oil Emulsion. *J. Dispers. Sci. Technol.* **2012**, *33*, 775–785. [CrossRef]
23. Dalmazzone, C.; Noïk, C.; Komunjer, L. Mechanism of crude-oil/water interface destabilization by silicone demulsifiers. *Spe J.* **2005**, *10*, 44–53. [CrossRef]
24. Zhang, L.; Ying, H.; Yan, S.; Zhan, N.; Guo, Y.; Fang, W. Hyperbranched poly(amido amine) demulsifiers with ethylenediamine/1,3-propanediamine as an initiator for oil-in-water emulsions with microdroplets. *Fuel* **2018**, *226*, 381–388. [CrossRef]
25. Hao, L.; Jiang, B.; Zhang, L.; Yang, H.; Sun, Y.; Wang, B.; Yang, N. Efficient Demulsification of Diesel-in-Water Emulsions by Different Structural Dendrimer-Based Demulsifiers. *Ind. Eng. Chem. Res.* **2016**, *55*, 1748–1759. [CrossRef]
26. Jabbari, M.; Izadmanesh, Y.; Ghavidel, H. Synthesis of ionic liquids as novel emulsifier and demulsifiers. *J. Mol. Liq.* **2019**, *293*, 111512. [CrossRef]
27. Zolfaghari, R.; Fakhru'l-Razi, A.; Abdullah, L.C.; Elnashaie, S.; Pendashteh, A. Demulsification techniques of water-in-oil and oil-in-water emulsions in petroleum industry. *Sep. Purif. Technol.* **2016**, *170*, 377–407. [CrossRef]
28. Liu, J.; Wang, H.; Li, X.; Jia, W.; Zhao, Y.; Ren, S. Recyclable magnetic graphene oxide for rapid and efficient demulsification of crude oil-in-water emulsion. *Fuel* **2017**, *189*, 79–87. [CrossRef]
29. Nikkhah, M.; Tohidian, T.; Rahimpour, M.R.; Jahanmiri, A. Efficient demulsification of water-in-oil emulsion by a novel nano-titania modified chemical demulsifier. *Chem. Eng. Res. Des.* **2015**, *94*, 164–172. [CrossRef]
30. Zhang, J.; Li, Y.; Bao, M.; Yang, X.; Wang, Z. Facile Fabrication of Cyclodextrin-Modified Magnetic Particles for Effective Demulsification from Various Types of Emulsions. *Environ. Sci. Technol.* **2016**, *50*, 8809–8816. [CrossRef]
31. Liu, J.; Li, X.; Jia, W.; Li, Z.; Zhao, Y.; Ren, S. Demulsification of Crude Oil-in-Water Emulsions Driven by Graphene Oxide Nanosheets. *Energy Fuels* **2015**, *29*, 4644–4653. [CrossRef]
32. Wang, H.; Liu, J.; Xu, H.; Ma, Z.; Jia, W.; Ren, S. Demulsification of heavy oil-in-water emulsions by reduced graphene oxide nanosheets. *RSC Adv.* **2016**, *6*, 106297–106307. [CrossRef]
33. Xu, H.; Jia, W.; Ren, S.; Wang, J.; Yang, S. Stable and efficient demulsifier of functional fluorinated graphene for oil separation from emulsified oily wastewaters. *J. Taiwan Inst. Chem. Eng.* **2018**, *93*, 492–499. [CrossRef]
34. McCoy, T.M.; Turpin, G.; Teo, B.M.; Tabor, R.F. Graphene oxide: A surfactant or particle? *Curr. Opin. Colloid Interface Sci.* **2019**, *39*, 98–109. [CrossRef]
35. Ma, M.; Xu, M.; Liu, S. Surface Chemical Modifications of Graphene Oxide and Interaction Mechanisms at the Nano-Bio Interface. *Acta Chim. Sin.* **2020**, *78*, 877–887. [CrossRef]
36. Wu, H.; Yi, W.; Chen, Z.; Wang, H.; Du, Q. Janus graphene oxide nanosheets prepared via Pickering emulsion template. *Carbon* **2015**, *93*, 473–483. [CrossRef]
37. Contreras Ortiz, S.N.; Cabanzo, R.; Mejía-Ospino, E. Crude oil/water emulsion separation using graphene oxide and amine-modified graphene oxide particles. *Fuel* **2019**, *240*, 162–168. [CrossRef]
38. Luo, D.; Zhang, F.; Zheng, H.; Ren, Z.; Jiang, L.; Ren, Z. Electrostatic-attraction-induced high internal phase emulsion for large-scale synthesis of amphiphilic Janus nanosheets. *Chem. Commun.* **2019**, *55*, 1318–1321. [CrossRef]
39. Herdes, C.; Ervik, Å.; Mejía, A.; Müller, E.A. Prediction of the water/oil interfacial tension from molecular simulations using the coarse-grained SAFT-γ Mie force field. *Fluid Phase Equilibria* **2018**, *476*, 9–15. [CrossRef]
40. Mohammed, S.; Mansoori, G.A. The Role of Supercritical/Dense CO_2 Gas in Altering Aqueous/Oil Interfacial Properties: A Molecular Dynamics Study. *Energy Fuels* **2018**, *32*, 2095–2103. [CrossRef]
41. Stephan, S.; Hasse, H. Interfacial properties of binary mixtures of simple fluids and their relation to the phase diagram. *Phys. Chem. Chem. Phys.* **2020**, *22*, 12544–12564. [CrossRef]
42. Nagl, R.; Zeiner, T.; Zimmermann, P. Interfacial Mass Transfer in Quaternary Liquid-Liquid Systems. *Chem. Eng. Process. Process Intensif.* **2022**, *171*, 108501. [CrossRef]
43. Stephan, S.; Schaefer, D.; Langenbach, K.; Hasse, H. Mass transfer through vapour–liquid interfaces: A molecular dynamics simulation study. *Mol. Phys.* **2020**, *119*, e1810798. [CrossRef]
44. Baidakov, V.G.; Protsenko, S.P. Molecular-Dynamics Simulation of Relaxation Processes at Liquid–Gas Interfaces in Single- and Two-Component Lennard-Jones Systems. *Colloid J.* **2019**, *81*, 491–500. [CrossRef]
45. Braga, C.; Muscatello, J.; Lau, G.; Muller, E.A.; Jackson, G. Nonequilibrium study of the intrinsic free-energy profile across a liquid-vapour interface. *J. Chem. Phys.* **2016**, *144*, 044703. [CrossRef]

46. Stephan, S.; Hasse, H. Enrichment at vapour–liquid interfaces of mixtures: Establishing a link between nanoscopic and macroscopic properties. *Int. Rev. Phys. Chem.* **2020**, *39*, 319–349. [CrossRef]
47. Stephan, S.; Becker, S.; Langenbach, K.; Hasse, H. Vapor-liquid interfacial properties of the system cyclohexane + CO_2: Experiments, molecular simulation and density gradient theory. *Fluid Phase Equilibria* **2020**, *518*, 112583. [CrossRef]
48. Chakraborty, S.; Ge, H.; Qiao, L. Molecular Dynamics Simulations of Vapor-Liquid Interface Properties of n-Heptane/Nitrogen at Subcritical and Transcritical Conditions. *J. Phys. Chem B* **2021**, *125*, 6968–6985. [CrossRef]
49. Lian, P.; Jia, H.; Wei, X.; Han, Y.; Wang, Q.; Dai, J.; Wang, D.; Wang, S.; Tian, Z.; Yan, H. Effects of zwitterionic surfactant adsorption on the component distribution in the crude oil droplet: A molecular simulation study. *Fuel* **2021**, *283*, 119252. [CrossRef]
50. Liu, X.; Li, Y.; Tian, S.; Yan, H. Molecular Dynamics Simulation of Emulsification/Demulsification with a Gas Switchable Surfactant. *J. Phys. Chem. C* **2019**, *123*, 25246–25254. [CrossRef]
51. Lucassen-Reynders, E.H.; Lucassen, J. Surface dilational viscosity and energy dissipation. *Colloids Surf. A Physicochem. Eng. Asp.* **1994**, *85*, 211–219. [CrossRef]
52. Falls, A.H.; Scriven, L.E.; Davis, H.T. Adsorption, structure, and stress in binary interfaces. *J. Chem. Phys.* **1983**, *78*, 7300–7317. [CrossRef]
53. Boek, E.S.; Yakovlev, D.S.; Headen, T.F. Quantitative Molecular Representation of Asphaltenes and Molecular Dynamics Simulation of Their Aggregation. *Energy Fuels* **2009**, *23*, 1209–1219. [CrossRef]
54. Khalaf, M.H.; Mansoori, G.A. A new insight into asphaltenes aggregation onset at molecular level in crude oil (an MD simulation study). *J. Pet. Sci. Eng.* **2018**, *162*, 244–250. [CrossRef]
55. Luo, D.; Wang, F.; Zhu, J.; Cao, F.; Liu, Y.; Li, X.; Willson, R.C.; Yang, Z.; Chu, C.-W.; Ren, Z. Nanofluid of graphene-based amphiphilic Janus nanosheets for tertiary or enhanced oil recovery: High performance at low concentration. *Proc. Natl. Acad. Sci. USA* **2016**, *113*, 7711–7716. [CrossRef] [PubMed]
56. Robertson, M.J.; Tirado-Rives, J.; Jorgensen, W.L. Improved Peptide and Protein Torsional Energetics with the OPLS-AA Force Field. *J. Chem. Theory Comput.* **2015**, *11*, 3499–3509. [CrossRef]
57. Dodda, L.S.; de Vaca, I.C.; Tirado-Rives, J.; Jorgensen, W.L. LigParGen web server: An automatic OPLS-AA parameter generator for organic ligands. *Nucleic Acids Res.* **2017**, *45*, W331–W336. [CrossRef]
58. Berendsen, H.J.C.P.; Postma, J.; Gunsteren, W.; Dinola, A.D.; Haak, J.R. Molecular-Dynamics with Coupling to An External Bath. *J. Chem. Phys.* **1984**, *81*, 3684. [CrossRef]
59. Bussi, G.; Donadio, D.; Parrinello, M. COMP 8-Canonical sampling through velocity rescaling. *Abstr. Pap. Am. Chem. Soc.* **2007**, *234*, 014101.
60. Parrinello, M.; Rahman, A. Polymorphic transitions in single crystals: A new molecular dynamics method. *J. Appl. Phys.* **1998**, *52*, 7182–7190. [CrossRef]
61. Essmann, U.; Perera, L.; Berkowitz, M.L.; Darden, T.; Lee, H.; Pedersen, L.G. A smooth particle mesh Ewald method. *J. Chem. Phys.* **1995**, *103*, 8577–8593. [CrossRef]
62. Hess, B. P-LINCS: A parallel linear constraint solver for molecular simulation. *J. Chem. Theory Comput.* **2008**, *4*, 116–122. [CrossRef] [PubMed]
63. Humphrey, W.; Dalke, A.; Schulten, K. VMD: Visual molecular dynamics. *J. Mol. Graph.* **1996**, *14*, 33–38. [CrossRef]
64. Liu, J.; Zhao, Y.P.; Ren, S.L. Molecular Dynamics Simulation of Self-Aggregation of Asphaltenes at an Oil/Water Interface: Formation and Destruction of the Asphaltene Protective Film. *Energy Fuels* **2015**, *29*, 1233–1242. [CrossRef]
65. Dreyer, D.R.; Park, S.; Bielawski, C.W.; Ruoff, R.S. The chemistry of graphene oxide. *Chem. Soc. Rev.* **2010**, *39*, 228–240. [CrossRef]

Article

Screening and Demulsification Mechanism of Fluorinated Demulsifier Based on Molecular Dynamics Simulation

Xiaoheng Geng [1,2,*], Changjun Li [1], Lin Zhang [3], Haiying Guo [2], Changqing Shan [2], Xinlei Jia [2,3], Lixin Wei [3], Yinghui Cai [4] and Lixia Han [4]

1. College of Petroleum Engineering, Southwest Petroleum University, Sichuan 610500, China; lichangjunemail@sina.com
2. College of Chemical Engineering and Safety, Binzhou University, Binzhou 256600, China; guohaiying1987@126.com (H.G.); sdscq@163.com (C.S.); 18434362466@126.com (X.J.)
3. School of Petroleum Engineering, Northeast Petroleum University, Daqing 163318, China; zl980110@163.com (L.Z.); weilixin73@163.com (L.W.)
4. Chambroad Chemical Industry Research Institute Co., Ltd., Binzhou 256505, China; yinghui.cai@chambroad.com (Y.C.); lixia.han@chambroad.com (L.H.)
* Correspondence: 052710@163.com

Abstract: In order to solve the problem of demulsification difficulties in Liaohe Oilfield, 24 kinds of demulsifiers were screened by using the interface generation energy (IFE) module in the molecular dynamics simulation software Materials Studio to determine the ability of demulsifier molecules to reduce the total energy of the oil–water interface after entering the oil–water interface. Neural network analysis (NNA) and genetic function approximation (GFA) were used as technical means to predict the demulsification effect of the Liaohe crude oil demulsifier. The simulation results show that the SDJ9927 demulsifier with ethylene oxide (EO) and propylene oxide (PO) values of 21 (EO) and 44 (PO) reduced the total energy and interfacial tension of the oil–water interface to the greatest extent, and the interfacial formation energy reached −640.48 Kcal/mol. NNA predicted that the water removal amount of the SDJ9927 demulsifier was 7.21 mL, with an overall error of less than 1.83. GFA predicted that the water removal amount of the SDJ9927 demulsifier was 7.41mL, with an overall error of less than 0.9. The predicted results are consistent with the experimental screening results. SDJ9927 had the highest water removal rate and the best demulsification effect. NNA and GFA had high correlation coefficients, and their R^2s were 0.802 and 0.861, respectively. The higher R^2 was, the more accurate the prediction accuracy was. Finally, the demulsification mechanism of the interfacial film breaking due to the collision of fluorinated polyether demulsifiers was studied. It was found that the carbon–fluorine chain had high surface activity and high stability, which could protect the carbon–carbon bond in the demulsifier molecules to ensure that there was no re-emulsion due to the stirring external force.

Keywords: demulsifier; fluorinated; demulsification mechanism; molecular dynamics simulation; neural network analysis; genetic function approximation

1. Introduction

Liaohe Oilfield in China mainly produces heavy oil and super heavy oil. It is very difficult to demulsify the oil emulsion because of its large asphaltene content and high viscosity [1]. Liaohe Oilfield has entered the middle and late stage of exploitation. Since water flooding cannot meet the demand of the oilfield production increase, alkali–surfactant–polymer (ASP) flooding technology can better meet the demand of the production increase in Liaohe Oilfield in the middle and later stages of production [2,3]. ASP flooding technology can greatly improve oil recovery, but the synergistic effect and emulsifying effect in the process of oil displacement make the emulsification of produced fluid very serious [4,5]. When the crude oil is produced by ASP flooding, a large number of surfactants, polymers and other chemicals are used, resulting in more complex produced fluid systems and more difficult demulsification [6]. Transportation and refining of this stable emulsion without treatment can cause serious problems, such as pipeline corrosion, scaling, increased equipment load and fuel consumption [7,8]. Conventional demulsification technology mainly includes physical demulsification, chemical demulsification and biological demulsification [9–11]. Chemical demulsification requires simpler equipment and has a lower cost and better demulsification effect. It can be used alone or combined with other demulsification methods to achieve efficient demulsification [12,13]. However, at present, many demulsifiers cannot meet the actual needs of Liaohe Oilfield or are difficult to apply due to cost, safety and other factors [14]. Therefore, the synthesis of a kind of demulsifier which is suitable for the development of Liaohe Oilfield and can efficiently and rapidly demulsify has become an urgent problem, which is of great significance to the good and efficient development of Liaohe Oilfield.

In recent years, molecular dynamics simulation technology has developed rapidly and is gradually applied to surfactants such as demulsifiers. Molecular dynamics simulation refers to the use of computer technology, discusses the interfacial structure and interfacial action of emulsion after adding the demulsifier at the molecular level in order to explain the role of demulsifiers through this technology, optimizes the selection of efficient demulsifiers and better serves oilfield production [15,16]. Using molecular dynamics simulation to guide experimental research not only makes the experimental data and their universal mechanism more visible, but also provides a new direction for future experimental research.

Marquez et al. [17] first studied the demulsification behavior of demulsifiers at the oil–water interfacial film of oil-in-water emulsion using an atomic model. They found that surfactants that can be used as demulsifiers must have the following characteristics: Firstly, the solubility of demulsifiers in the aqueous phase must be higher than that in the oil phase. Secondly, they must have certain diffusions and concentrations. Finally, the surface activity of demulsifiers must be higher than that of emulsifiers. The demulsifier with the above characteristics can reach the oil–water interface film and reduce the stability of the interface film to achieve demulsification.

Ballal et al. [18] used the improved iSAFT (interfacial statistical association fluid theory) to explore the influence of poly (ethylene oxide)–propylene oxide polyether on the interfacial film of water–toluene by studying the molecular weight, the ratio of EO to PO, branching degree and order degree, so as to understand the influence of demulsifier structure on the interfacial film at the molecular level and predict the performance of real demulsifiers. The results show that the interfacial tension decreased with the increase in molecular weight and the number of branched chains. When EO:PO = 1:1, the interfacial tension is at its minimum. Moreover, the surface activity of PEO-PPO-PEO is higher than that of PPO-PEO-PPO.

Zhang et al. [19] used a polyamide-amine dendrimer demulsifier to study the effect of the hydrophobic chain on interfacial properties and demulsification with molecular dynamics simulation technology. The results show that with the increase in the demulsifier concentration, the kinetic parameters n and t* obtained by characterizing the molecular diffusion rate decreased. At the same time, unlike the traditional demulsifier adsorption and diffusion behavior, with the increase in the hydrophobic chain length, the t* value decreased and the n value increased, showing a slow diffusion–adsorption process.

A machine model algorithm can predict and integrate new rules and development trends from a large number of data texts in multiple dimensions. In general, the process of using machine algorithm to simulate experimental data can be divided into two steps: inputting old data and simulating new trends. With the development and widespread application of computer algorithms, researchers often use the neural network algorithm and genetic algorithm to predict the mixed-phase pressure, and good prediction results have been achieved.

The purpose of this study was to provide efficient and economical fluorinated polyether demulsifiers for Liaohe Oilfield. Compared with general demulsifiers, the fluorine atoms contained in this demulsifier can partially or completely replace the hydrogen atoms on the hydrocarbon chain, so that the nonpolar groups in the demulsifier can form carbon–fluorine bonds with stronger bond energy, and this carbon–fluorine chain with higher bond energy can show strong stability. Fluorinated polyether demulsifiers have better surface activity, chemical stability, thermal stability and compatibility than conventional demulsifiers. Fluorinated hydrocarbon groups are also hydrophobic, which can reduce pollution. The interfacial generation energy (IFE) in molecular dynamics was used to screen 24 kinds of demulsifiers. Neural network analysis (NNA) and genetic function approximation (GFA) were applied to predict demulsification, so as to look for the rules from the existing experimental data to obtain the corresponding prediction conclusions.

2. Experimental

2.1. Materials

Tetraethylenepentaamine was purchased from Beijing Tianyu Kanghong Chemical Technology Co., Ltd. (Beijing, China). P-trifluoromethyl phenol was purchased from Shanghai Sahn Chemical Technology Co., Ltd. (Shanghai, China). Formaldehyde was purchased from Shanghai Macklin Biochemical Technology Co., Ltd. (Shanghai, China). Xylene and toluene were ordered from Shanghai Jizhi Biochemical Technology Co., Ltd. (Shanghai, China). Potassium hydroxide was purchased from Shanghai Sibaiquan Chemical Co., Ltd. (Shanghai, China). Potassium hydroxide was purchased from Shanghai Sibaiquan Chemical Co., Ltd. Ethylene oxide (EO) and propylene oxide (PO) were purchased from Zibo Shandong Zixiang Sales Chemical Co., Ltd. (Zibo, China). The tested oil sample was produced from fluid from a block in Liaohe Oilfield. Fluorinated demulsifiers were synthesized by using trifluoromethyl phenol, formaldehyde and other raw materials as initiators and then synthesized through polymerization reaction with propylene oxide and ethylene oxide [20]. The physiochemical characteristics are shown in Table 1.

Table 1. Basic physical properties of crude oil produced from a block from Liaohe Oilfield.

Density $kg \cdot m^{-3}$	Dynamic Viscosity (50 °C) $mPa \cdot s$	Gum %	Asphaltene %	Acid Value $mgKOH \cdot g^{-1}$	Pour Point °C	Moisture Content %	Saturate %	Aromatic %	Resin %
943.0	180.2	24.93	12.75	2.45	17.3	17.02	46.63	15.69	32.83

2.2. Molecular Optimization and Model Construction

All the simulations were performed on the molecular dynamics software Materials Studio2018. The interaction parameters of surfactants came from the condensed-state optimized molecular force field—COMPASS force field.

Firstly, the 3D model structures of n-decane and demulsifier molecules were built by using the Visualizer module in the program, and the geometric optimization of the structures of the three surfactant molecules was carried out by using the Smart method through the COMPASS force field of the Dmol3 module, so that the surface molecular system could achieve the minimum energy, and the optimized molecular structure of the optimal molecular conformation was obtained, as shown in Figure 1.

Then, the crude oil system model and demulsifier system model were established at 278 K by using the construction tool under AC module, COMPASS force field and Periodic Cell periodic boundary conditions. Based on the position reference of the rectangular coordinate system, the size of the system box was set. With the origin as the center, the lengths in x, y and z directions were 4 nm × 4 nm × 12 nm, respectively. The system model is shown in Figure 2. The simulation systems with different EO/PO ratios were composed of 2000 n-sunane molecules and 500 water molecules.

Finally, the Dynamics tool under the Forcite module was used. The simulation level was MEDIUM, and the simulation system ensemble was NVT ensemble, keeping the system at a constant temperature of 298 K. AtomBased was used to represent van der Waals interaction and electrostatic interaction. Andersen method was selected for parameter control of ensemble, namely temperature control. In addition, Berendsen method was used for pressure change. A 3000-step process was established and the last nanosecond result was obtained by statistical analysis.

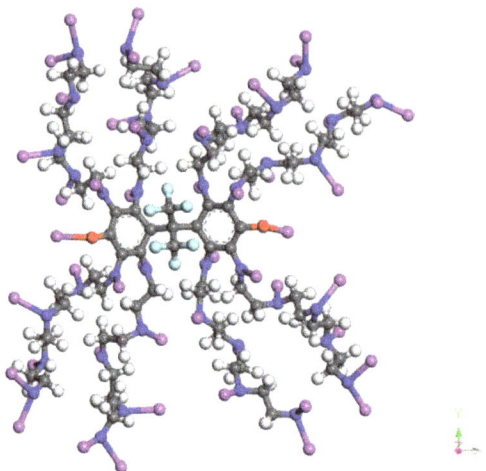

Figure 1. Schematic diagram of optimized molecular structure, where blue is N, red is O, white is H, gray is C, and purple is $R_3 = (C_3H_6O)_x(C_2H_4O)_y$.

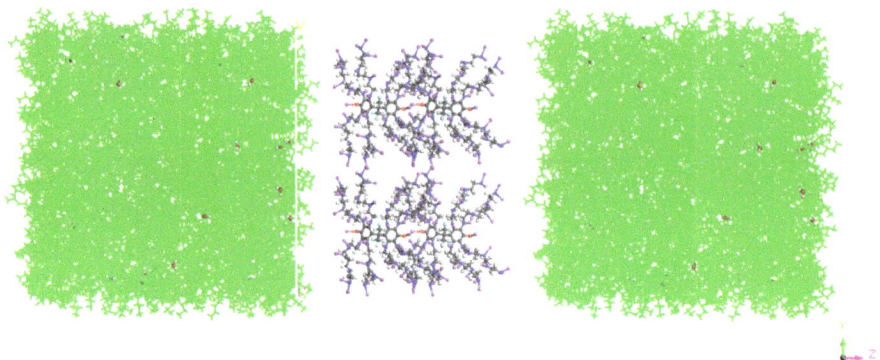

Figure 2. Demulsifier system model.

2.3. Neural Network Analysis (NNA)

Neural network analysis (NNA) refers to the method of machine learning and data processing that controls various parameters and layers based on the way in which neurons in the brain of organisms transmit information [21,22]. NNA was first proposed by McCulloch and Pitts in 1943 as a way to simulate the analysis of neurons in the brain. Although it makes too many assumptions and simplifications than real brain neurons, it still contributes considerable intelligence in research. Therefore, NNA has considerable research and application value. Since then, NNA has been greatly developed, and hundreds of models have been proposed. Figure 3 is a complete typical three-layer neural network structure, which is divided into three parts: multinode input layer, single-node output layer and hidden layer. A three-layer BP neural network can solve almost all the prediction problems near exact precision, so only one hidden layer was used in this study.

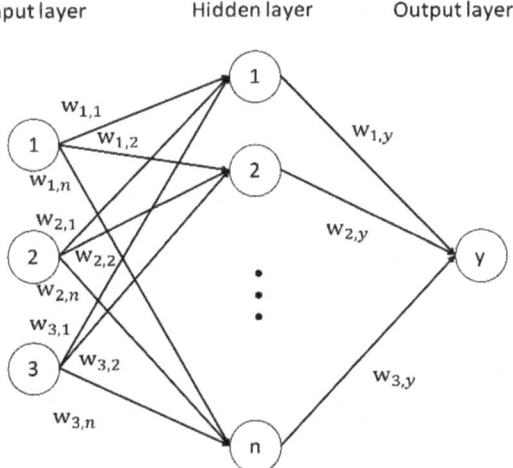

Figure 3. Basic structure of three-layer neural network.

Neurons refer to nodes, which are the most basic structure of neural networks. Each node is an information-processing element. In addition to the input layer, each node uses the transformed linear combination of node output from the lower layer as its input:

$$I_i = \sum_j w_{ij} X_j + \theta_i \qquad (1)$$

In the expression, I_i is the input to the i node, X_j is the output of the j node in the previous layer, j is the summation of all nodes in the previous layer, w_{ij} is the connection weight between nodes, and θ_i is the offset value. It is worth noting that the nodes between each layer are not fixed, which needs to be set according to the actual situation. When exploring MMP, the nodes in the input layer are designed as multiple variables that affect MMP, and the output is the corresponding MMP value. The number of hidden layers and the number of nodes in each layer can be defined by users themselves, so as to make the output results closer to the real value.

The transfer function needs to be realized through the transfer function between the input layer and the hidden layer and between the hidden layer and the output layer. In addition, the conversion information is realized by setting the weights and bias values. In this way, the data between the input layer and the output layer can be connected and their relationship can be directly reflected. The study uses a transfer function called Sigmoid transfer function (Formula (2)), which allows for easy differentiation and has a smooth function to achieve data output in a narrow range.

$$y = \frac{1.2}{1 + e^{-x}} - 0.1 \qquad (2)$$

After setting the above information, training should be started and the neural network should be learned independently. The training minimizes the error and makes the final prediction more accurate. Here, BFGS algorithm is used to find the minimum value of the error, and the error function (3) is used to determine the matching degree between the calculated output and the expected output:

$$E = \sum_{i=1}^{n} C_i(y_i - y_i')^2 + Q\sum_j (x_j - \overline{x_j})^2 + P\sum_{k,l} w_{k,l}^2 \qquad (3)$$

where C_i is the parameter value of the proportion of results. In this study, the item is 1, y_i and y'_i represent the true value and the predicted value, respectively. Q is the penalty factor set for the missing value, x_j is the missing data value of the system guess, $\overline{x_j}$ is the average value of each input data, P represents the penalty factor of connection weight, and $w_{k,l}$ represents the connection weight. The first item is the main item of the error, which is the sum of squares of the difference between the predicted value and the actual output value of the model. The second item represents the error caused by filling the missing data. In addition, the average connection weight is added to the error function to prevent the collapse or error caused by excessive weight. In this way, the learning cycle iteration of the neural network can be carried out until the error drops to a certain level, and finally a trained neural network can be obtained.

2.4. Genetic Function Approximation (GFA)

Genetic function approximation (GFA) is carried out through selection, crossover and mutation [23,24]. This algorithm is based on Darwin's theory of evolution and some viewpoints of genetics. Generally speaking, choosing an excellent father will lead to better offspring. In addition, the mutation operation accelerates the progress of the algorithm and does not fall into local optimum [25]. By applying this algorithm to the field of intelligence, the optimal selection method is obtained to improve the economic benefits. The structural process is shown in Figure 4.

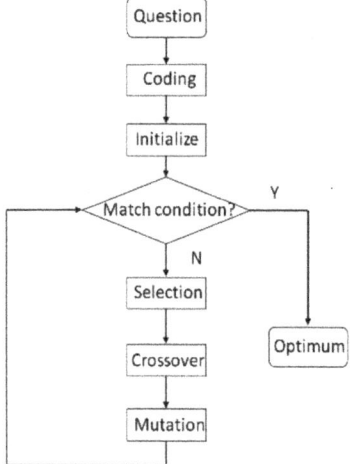

Figure 4. Genetic function approximation chart.

The parallel method can improve the convergence speed under the premise of ensuring a certain fitness, so as to obtain more accurate results. The method extended by this method is the genetic function approximation method. This method has good performance, including high robustness, and can be used to improve fault tolerance.Therefore, this method has been widely used in practical applications.

The core idea of GFA is similar to that of GA (genetic algorithm), which encodes the region to be searched as one or more strings, each string representing a position in the search space, where each group of strings is called a population, and the evolution of the population makes it move towards the search target. At the beginning of the setting, the initial clock group is set to 100, and through 500 iterations, in the iterative process, there is a choice of crossover and mutation. The members through variation need to score; namely, the following fitness function is used to score (Formula (4)). In GFA, the evaluation criteria of the model are related to the quality of data regression fitting. The fitness function used in this study was Friedman LOF function:

$$F = \frac{SSE}{M[1 - \lambda(\frac{C+dp}{M})]^2} \quad (4)$$

where SSE is the sum of squares of errors, C is the number of items in the model, which is not a constant, p is the total number of descriptors contained in all model items, M is the number of samples in the training set, λ is a safety factor, the value is 0.99, to ensure that the denominator of the expression does not become zero, and d is the scaling smoothing parameter; the following expression is associated with the specified scaling LOF smoothing parameter:

$$d = \alpha(\frac{M - C_{max}}{C_{max}}) \quad (5)$$

C_{max} means the maximum equation length.

Selection is not random; individuals with good adaptability are chosen. After the reproduction of the selected individuals, the new members need to be graded to determine whether they are the next selected object. Crossing process is the exchange of genetic information between parent chromosomes. The selection of genetic information in the cross process is random. When the cross selection is over, the new members need to be graded to determine whether they are remixed into the population to seek better results.

Parents:

$$x_1^2, x_2 \big| x_4, x_3^2 \quad (6)$$

$$x_1, x_3 \big| x_4, x_5^2 \quad (7)$$

Child:

$$x_1^2, x_2 \big| x_4, x_5^2 \quad (8)$$

In order to make genetics more scientific, mutation operation is needed. What is reflected in the computer is the mutual transformation between 0 and 1 so as to find the optimal solution faster.

Finally, through a series of genetic operations, the offspring are optimized, and the mutation operation is used to prevent the calculation results from falling into local optimum, leading to wrong results. The optimal population obtained by this method was the optimal solution we found

2.5. Turbidity Point and HLB Value Test

The cloud point of fluorinated polyether demulsifier was determined by Cintra 10e UV-Vis spectrometer (GBC Scientific Instruments Company, Melbourne, Australia). The HLB (hydrophilic–lipophilic balance) value of demulsifier was calculated according to the cloud point of surfactant and the empirical formula of HLB to obtain the corresponding HLB value. The empirical formula is as follows:

$$HLB = 0.0980X + 4.02 \quad (9)$$

X is the cloud point value of 1wt% fluorinated polyether demulsifier.

2.6. Experiment on Demulsification and Water Removal of Demulsifier

Crude oil emulsion with water content of 17.02% was placed into constant-temperature water bath heated to 55 °C for 30 min and then put into a stirring motor for 8 min at a speed of 2000 r/min. After that, it was put into the stirring machine for 5 min. A total of 50 mL crude oil emulsion was poured into a calibrated test tube and put it into a water bath heated to 60 °C and kept at a constant temperature for 25 min. Care was taken to ensure the height of the water surface did not exceed the height of the crude oil in the test tube.

The demulsifier was added into the test tube with micropipeter, and the cork was tightened. The test tube was turned upside down and shaken 3–5 times, and the cork was loosened to let off air. The bottle was recorked, and the tube was shaken 150 times by hand to fully mix the demulsifier and crude oil emulsion. After the cork was capped, the bottle was placed in a water bath at 60 °C for settling. The volume of dehydration at different times was observed to obtain the dehydration volume V. The blank sample without demulsifier addition was set to obtain the dehydration amount V_b. Therefore, the dehydration amount after adding demulsifier was $V_d = V - V_b$.

Demulsification efficiency is calculated as follows:

$$\text{Efficiency (\%)} = (V_O - V_d)/V_O \times 100$$

where V_O is the volume of water (water content) in the crude oil emulsion and V_d is the volume of water remaining in the oil phase after demulsifier addition.

3. Results and Discussion

3.1. Molecular Dynamics Simulation Results

The energy optimization trend and optimization steps of molecular dynamics simulation of some demulsifiers are shown in the Figures 5–18.

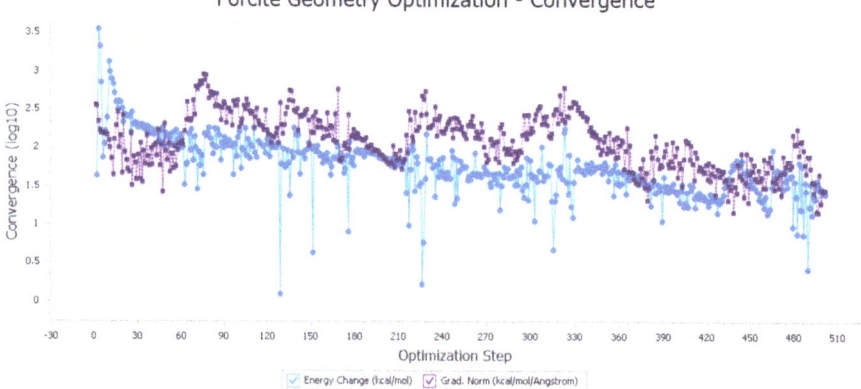

Figure 5. Demulsifier2 # oil–water interface geometric optimization steps.

Figure 6. Demulsifier2 # oil–water interface geometry optimization energy trend.

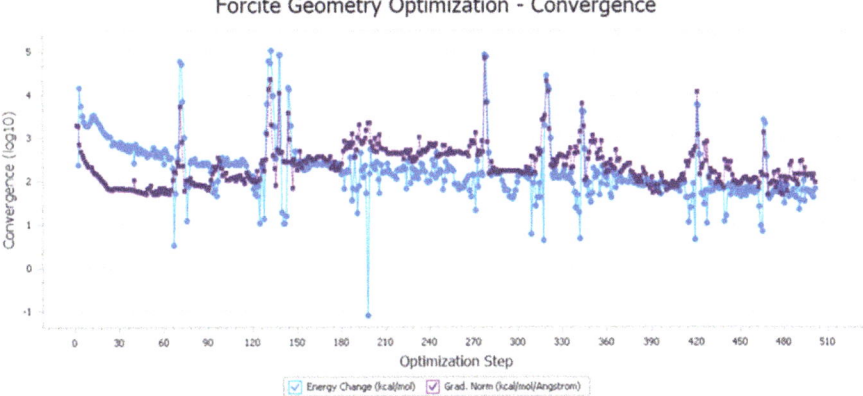

Figure 7. Demulsifier4 # oil–water interface geometric optimization steps.

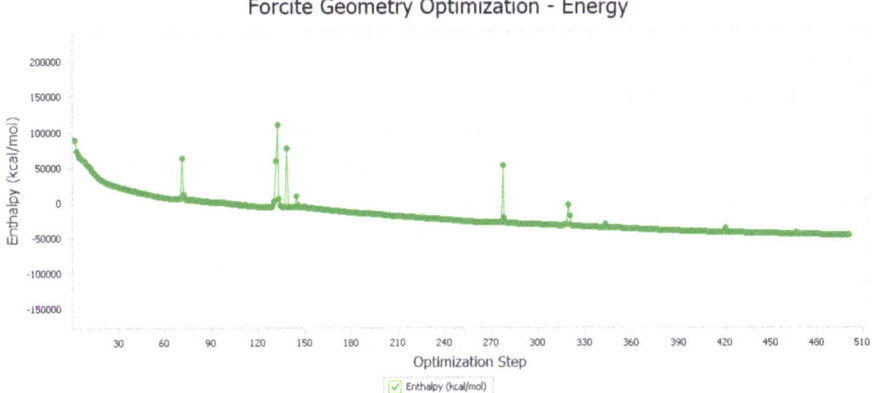

Figure 8. Demulsifier4 # oil–water interface geometry optimization energy trend.

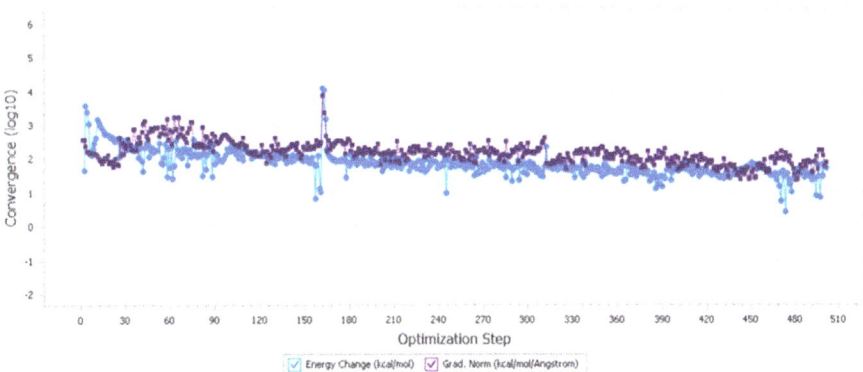

Figure 9. Demulsifier6 # oil–water interface geometric optimization steps.

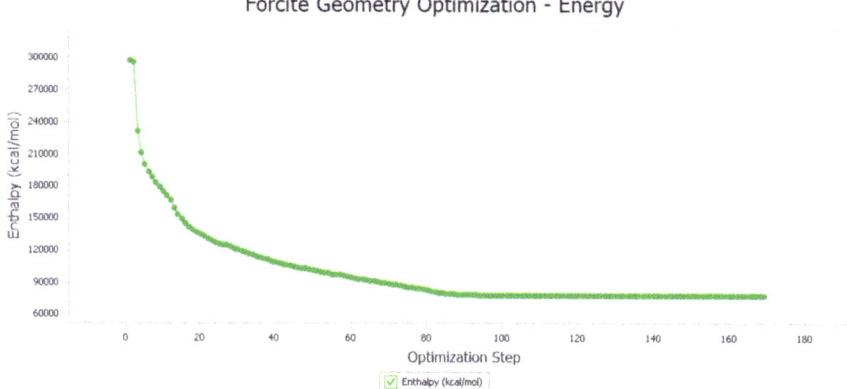

Figure 10. Demulsifier6 # oil–water interface geometry optimization energy trend.

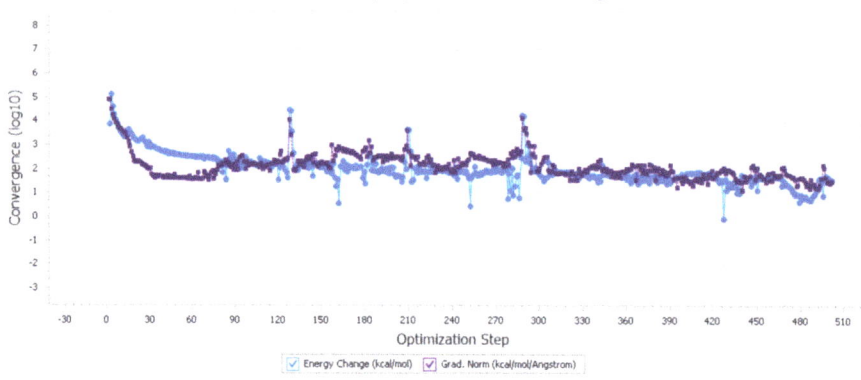

Figure 11. Demulsifier8 # oil–water interface geometric optimization steps.

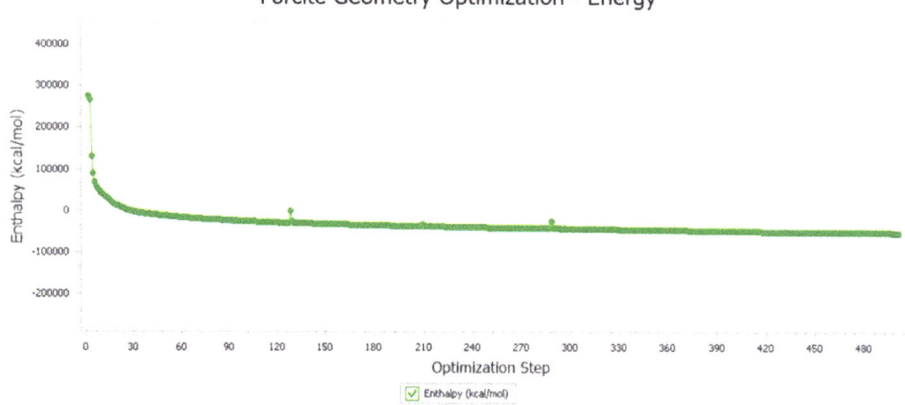

Figure 12. Demulsifier8 # oil–water interface geometry optimization energy trend.

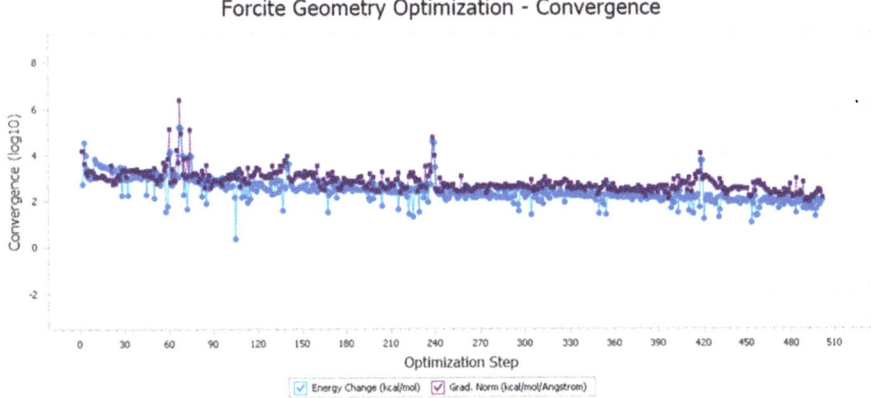

Figure 13. Demulsifier10# oil–water interface geometric optimization steps.

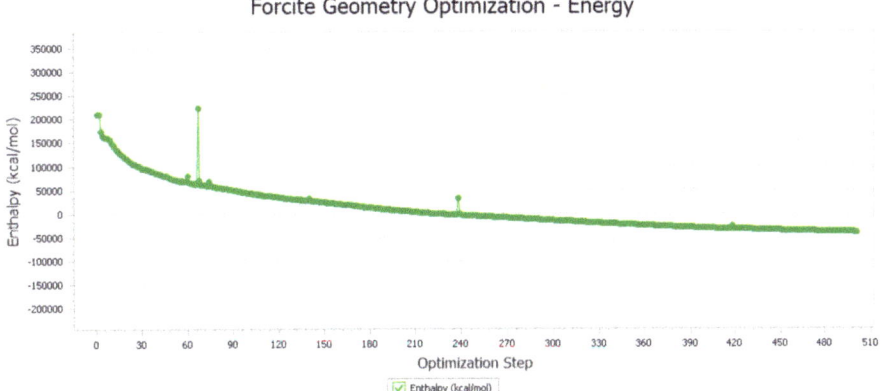

Figure 14. Demulsifier 10 # oil–water interface geometry optimization energy trend.

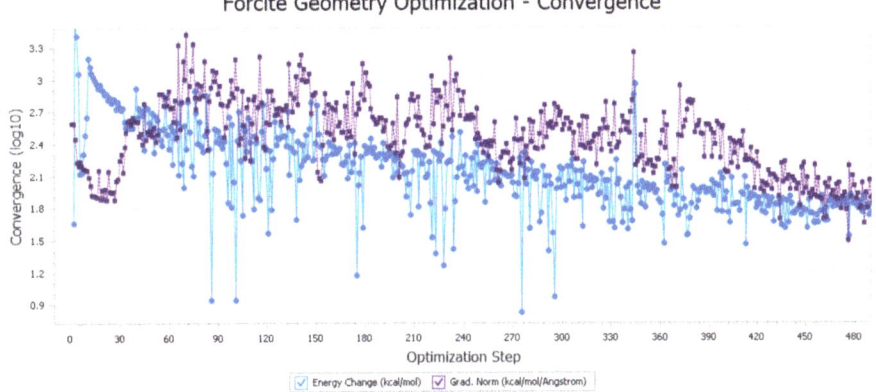

Figure 15. Demulsifier12# oil–water interface geometric optimization steps.

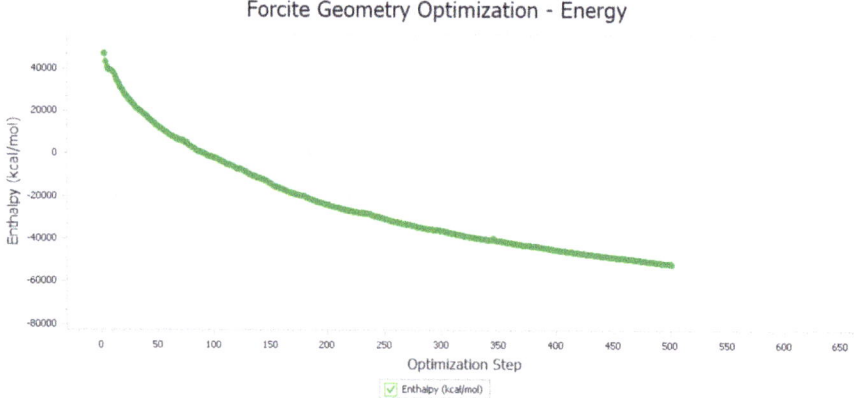

Figure 16. Demulsifier 12 # oil–water interface geometry optimization energy trend.

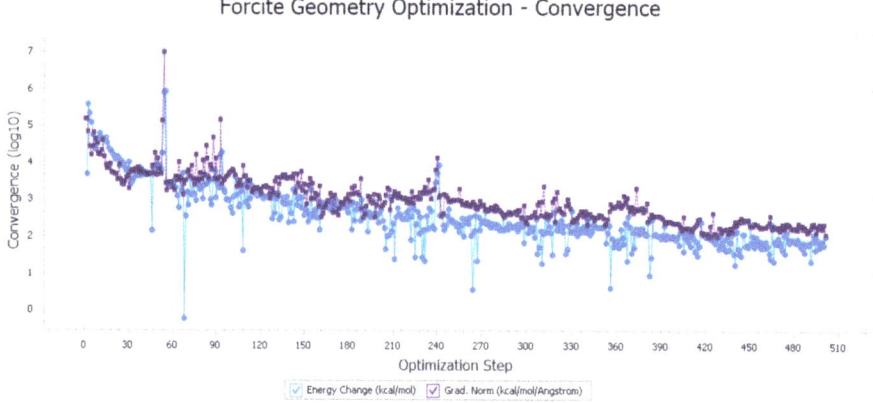

Figure 17. Demulsifier14# oil–water interface geometric optimization steps.

Figure 18. Demulsifier 14 # oil–water interface geometry optimization energy trend.

When the system was balanced, the free energy of the solution reached its minimum. Therefore, there was a chemical potential equilibrium relationship between surfactant monomer and micelle:

$$\mu_g^0 + KT \ln X_g = g(\mu_l^0 + KT \ln X_l) \tag{10}$$

where μ_l^0 is the standard chemical potential of surfactant monomer, X_l is the molar composition of the surfactant monomer, μ_g^0 is the standard chemical potential of micelles containing g surfactant monomers, and X_g is the molar composition of surfactant micelles.

Surfactants are dispersed in water in a molecular state, and their hydrophobic ends are arranged at the water interface to form a glacier structure, which reduces the entropy of the system [26]. However, when the hydrophobic end of the surfactant leaves the water interface, the surfactant molecules associate and the glacier structure is destroyed; then, the water molecules are separated from the bondage, thereby increasing the entropy of the system [27]. The formation process of micelles is a spontaneous entropy-driven process. In this process, the chaos of the system increases, and the total formation energy becomes negative [28].

The interfacial generation energy (IFE) refers to the reduced energy of the system after the surfactant molecules enter the oil–water interfacial layer. The stability of the interface can be investigated, and its value is closely related to the interaction force between surfactant and water molecules, surfactant molecules, surfactant and oil molecules. The calculation formula is as follows:

$$IFE = \frac{E_{total} - (n \times E + E_{ref})}{n} \tag{11}$$

where E_{total} is the total energy of the surfactant system after equilibrium at the oil–water interface, Kcal/mol. E_{ref} is the energy of oil–water interface system without the demulsifier, Kcal/mol (−42933.07). N is the number of demulsifier molecules in the system. E is the potential energy of the demulsifier molecule, Kcal/mol.

Table 2 shows that IFE value was negative, indicating that the energy of the whole system decreased. Therefore, after analyzing the absolute value of IFE, it was determined 6# demulsifier had the best effect.

Table 2. Interfacial generation energy of demulsifier.

Number	Determine of x,y	Water Removal/mL	Etotal/ (Kcal/mol)	IFE/ (Kcal/mol)
1#	SDJ6920: x = 31, y = 20	3.5	−47,819.55	−209.31
2#	SDJ6927: x = 31, y = 15	4	−47,869.13	−214.28
3#	SDJ6930: x = 31, y = 13	2.5	−47,002.77	−127.64
4#	SDJ6937: x = 31, y = 11	2	−46,933.29	−120.69
5#	SDJ9920: x = 44, y = 29	7	−51,139.1	−541.27
6#	SDJ9927: x = 44, y = 21	8	−52,131.12	−640.48
7#	SDJ9930: x = 44, y = 19	7.5	−51,943.12	−621.68
8#	SDJ9937: x = 44, y = 16	7	−51,611.02	−588.47
9#	SDJ15920: x = 71, y = 47	7	−51,002.13	−527.58
10#	SDJ15927: x = 71, y = 35	4	−47,133.09	−140.67
11#	SDJ15930: x = 71, y = 31	6	−50,169.33	−444.30
12#	SDJ15937: x = 71, y = 25	6	−50,129.32	−440.30
13#	SDJ19920: x = 89, y = 59	6	−50,196.02	−447.00
14#	SDJ19927: x = 89, y = 43	7	−51,584.18	−585.78
15#	SDJ19930: x = 89, y = 39	7	−50,113.22	−438.69
16#	SDJ19937: x = 89, y = 32	6.5	−51,008.12	−528.18
17#	SDJ29920: x = 134, y = 88	4	−47,003.72	−127.74
18#	SDJ29927: x = 134, y = 65	5.5	−48,121.12	−239.48
19#	SDJ29930: x = 134, y = 59	7	−51,341.91	−561.55
20#	SDJ29937: x = 134, y = 48	6.5	−51,620.12	−589.38
21#	SDJ39920: x = 178, y = 118	5.5	−49,723.18	−399.68
22#	SDJ39927: x = 179, y = 87	6	−50,339.33	−461.30
23#	SDJ39930: x = 179, y = 79	5	−48,013.13	−228.68
24#	SDJ39937: x = 17, y = 64	6	−50,632.44	−490.61

3.2. NNA and GFA Prediction Results

The prediction results of 24 demulsifiers are shown in Table 3. NNA and GFA were used to predict the demulsification effect of 24 demulsifiers, and the results are shown in Table 4. As shown in Table 3, where X and Y are PO and EO values, demulsification dehydration represents a demulsification effect.

Table 3. Cloud point and HLB value off fluorinated polyether demulsifiers.

Number	Determination of x, y Values	Cloud Point/°C	HLB Value
1#	SDJ6920: x = 31, y = 20	53.2	9.23
2#	SDJ6927: x = 31, y = 15	52.5	9.17
3#	SDJ6930: x = 31, y = 13	52.1	9.13
4#	SDJ6937: x = 31, y = 11	51.8	9.10
5#	SDJ9920: x = 44, y = 29	47.8	8.70
6#	SDJ9927: x = 44, y = 21	47.3	8.66
7#	SDJ9930: x = 44, y = 19	46.9	8.62
8#	SDJ9937: x = 44, y = 16	46.5	8.58
9#	SDJ15920: x = 71, y = 47	41.3	8.07
10#	SDJ15927: x = 71, y = 35	40.7	8.01
11#	SDJ15930: x = 71, y = 31	40.4	7.98
12#	SDJ15937: x = 71, y = 25	40.0	7.94
13#	SDJ19920: x = 89, y = 59	36.2	7.57
14#	SDJ19927: x = 89, y = 43	35.4	7.49
15#	SDJ19930: x = 89, y = 39	35.0	7.45
16#	SDJ19937: x = 89, y = 32	34.7	7.42
17#	SDJ29920: x = 134, y = 88	31.2	7.08
18#	SDJ29927: x = 134, y = 65	30.8	7.04
19#	SDJ29930: x = 134, y = 59	30.3	6.99
20#	SDJ29937: x = 134, y = 48	30.1	6.97
21#	SDJ39920: x = 179, y = 118	28.4	6.80
22#	SDJ39927: x = 179, y = 87	27.9	6.75
23#	SDJ39930: x = 179, y = 79	27.5	6.72
24#	SDJ39937: x = 179, y = 64	27.0	6.67

Table 4. The actual water removal amount of demulsifiers (120 min, 100 ppm).

Number	Determination of x, y Values	Water Removal Amount/mL	Demulsification Rate/%
1#	SDJ6920: x = 31, y = 20	3.5	41.13
2#	SDJ6927: x = 31, y = 15	4	47.00
3#	SDJ6930: x = 31, y = 13	2.5	29.38
4#	SDJ6937: x = 31, y = 11	2	23.50
5#	SDJ9920: x = 44, y = 29	7	82.26
6#	SDJ9927: x = 44, y = 21	8	94.01
7#	SDJ9930: x = 44, y = 19	7.5	88.13
8#	SDJ9937: x = 44, y = 16	7	82.26
9#	SDJ15920: x = 71, y = 47	7	82.26
10#	SDJ15927: x = 71, y = 35	4	47.00
11#	SDJ15930: x = 71, y = 31	6	70.51
12#	SDJ15937: x = 71, y = 25	6	70.51
13#	SDJ19920: x = 89, y = 59	6	70.51
14#	SDJ19927: x = 89, y = 43	7	82.26
15#	SDJ19930: x = 89, y = 39	7	82.26
16#	SDJ19937: x = 89, y = 32	6.5	76.38
17#	SDJ29920: x = 134, y = 88	4	47.00
18#	SDJ29927: x = 134, y = 65	5.5	64.63
19#	SDJ29930: x = 134, y = 59	7	82.26
20#	SDJ29937: x = 134, y = 48	6.5	76.38
21#	SDJ39920: x = 179, y = 118	5.5	64.63
22#	SDJ39927: x = 179, y = 87	6	70.51
23#	SDJ39930: x = 179, y = 79	5	58.75
24#	SDJ39937: x = 179, y = 64	6	70.51
Black	No demulsifier	0.3	3.53

Firstly, demulsification experiments were carried out for 24 kinds of demulsifiers to obtain the actual demulsification effect, and the results are shown in Table 3. Then, NNA and GFA were used to predict the demulsification effect of 24 demulsifiers, respectively, and the results are shown in Table 4. As shown in Table 3, X and Y are the values of PO and

EO, and the amount of water removal represents the demulsification effect. Figure 19 is the molecular structure of fluorinated demulsifier.

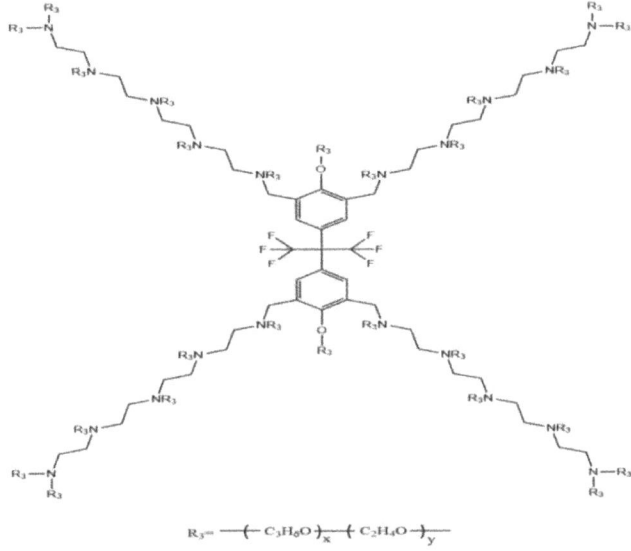

Figure 19. Molecular structure of fluorinated demulsifier.

The calculation results for the HLB value are shown in Table 3.

The values of x and y in R_3 are given in Table 4.

It can be seen from Figures 20 and 21 that GFA had a slightly higher correlation coefficient, but from Table 5, it can be found that the correlation coefficients of both were lower than 0.9, and both were above 0.8. In general, these two methods can predict the demulsification effect of this type of demulsifier. The square value of the correlation coefficient (R^2, coefficient of determination) reflects that the greater the R^2, the stronger the predictive ability.

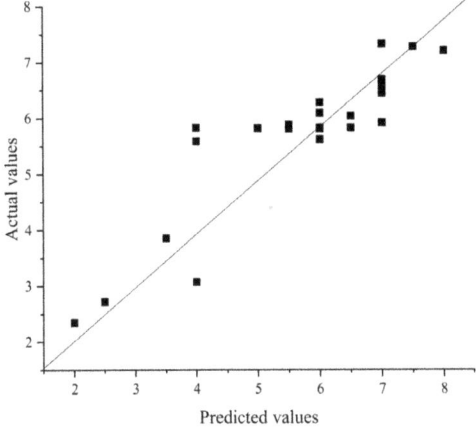

Figure 20. Comparison of predicted and actual values by NAA ($r^2 = 0.802$).

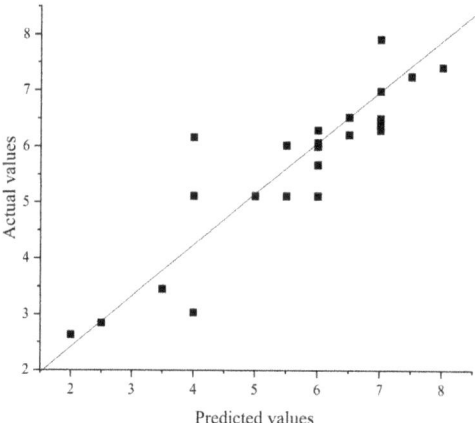

Figure 21. Comparison of predicted and actual values by GFA (r^2 = 0.861).

Table 5. Comparative analysis table between predicted and actual values.

Number	Actual Values	NNA Prediction	Differential Value	GFA Prediction	Differential Value
1#	3.5	3.85	−0.35	3.45	0.04
2#	4	3.07	0.93	3.03	0.97
3#	2.5	2.72	−0.22	2.84	−0.34
4#	2	2.35	−0.35	2.630	−0.630
5#	7	6.69	0.31	7.92	−0.92
6#	8	7.21	0.79	7.41	0.59
7#	7.5	7.28	0.22	7.25	0.25
8#	7	7.33	−0.33	6.99	0.01
9#	7	6.52	0.48	6.36	0.64
10#	4	5.59	−1.59	6.16	−2.16
11#	6	5.62	0.38	6.00	−0.00
12#	6	6.09	−0.09	5.67	0.33
13#	6	6.28	−0.28	6.29	−0.29
14#	7	6.58	0.42	6.50	0.50
15#	7	6.44	0.56	6.44	0.56
16#	6.5	5.83	0.67	6.21	0.29
17#	4	5.83	−1.83	5.11	−1.11
18#	5.5	5.88	−0.38	6.02	−0.52
19#	7	5.92	1.07	6.29	0.71
20#	6.5	6.04	0.46	6.52	−0.020
21#	5.5	5.81	−0.31	5.11	0.39
22#	6	5.81	0.18	5.11	0.89
23#	5	5.82	−0.82	5.11	−0.11
24#	6	5.83	0.17	6.070	−0.07

The GFA prediction formula can be edited into:

$$\text{Water removal amount} = 0.626504940 \times \text{RAMP}(65.555188173 - X) + 0.069567811 \times \text{RAMP}(78.604247624 - Y) - 0.012461417 \times [\text{RAMP}(74.580034129-X)]^2 - 0.001476625 \times [\text{RAMP}(70.127414082 - Y)]^2 + 5.110992651 \tag{12}$$

X and Y are the number of EOs and POs in the experiment, RAMP is the slope function.

With this prediction function, such molecules can be predicted through the formula, and the values of different proportions of X and Y and their dehydration rates can be roughly known, which can greatly save time.

3.3. Demulsification Mechanism

Substances that ensure oil–water phase dispersion and do not interfere with each other are called oil–water interfacial films [27]. The formation mechanism of an oil–water interface film is mainly as follows: Natural emulsifiers such as asphaltene and colloid in crude oil emulsion are stably adsorbed on the surface of water droplets, forming an interfacial film with low surface tension and interfacial free energy [28]. The demulsification mechanism of fluorinated polyether demulsifier in this study was the main mechanism of breaking the interface film. With the large-scale use of polymer demulsifiers, the mechanism of breaking interfacial film is increasingly recognized by a large number of researchers [29]. This kind of polymer surfactant has been favored by many oil fields because of its economy. When it is applied to specific crude oil demulsification, the dosage is very small, and the demulsification is very high [30].

The fluorinated polyether demulsifier developed in this paper is a nonpolar surfactant, which introduces a fluorine atom instead of a hydrogen atom to a hydrocarbon chain. The bond energy of the C-F bond is higher than that of the C-H bond, but the polarity is lower than that of the C-H bond [31,32]. Due to the characteristics of the fluorocarbon chain, compared with ordinary demulsifiers, it can reduce the oil–water interfacial tension more rapidly, accelerate the aggregation of water droplets and has better demulsification effect. The demulsification mechanism is essentially that surfactant molecules replace and break the interfacial film to release captured oil particles. Surfactants are added to the emulsion, and because of its higher interfacial activity, they replace the natural emulsifier molecules, such as asphaltene and colloid, adsorbed on the oil–water interfacial film and rearrange the oil–water interface, resulting in the rapid coalescence of water droplets and the realization of oil–water separation. The hydrophilicity of the hydrophilic block (PEO) of the block polyether demulsifier is higher than that of asphaltene molecules in oil. Therefore, the hydrophilic block (PEO) of the block polyether demulsifier can rapidly replace asphaltene molecules at the oil–water interface. When subjected to heating or shaking, the Brownian motion of macromolecules in the emulsion is intensified, and the number of collisions between macromolecules is increased. Therefore, the unstable interfacial film formed by demulsifier molecules is broken. Demulsifier molecules have higher stability because of the high bond energy of the C-F bond and the shielding property of the C-C, which ensures that there is no re-emulsion due to excessive stirring and other factors. Figure 22 depicts a diagram of the demulsification mechanism of the demulsifier.

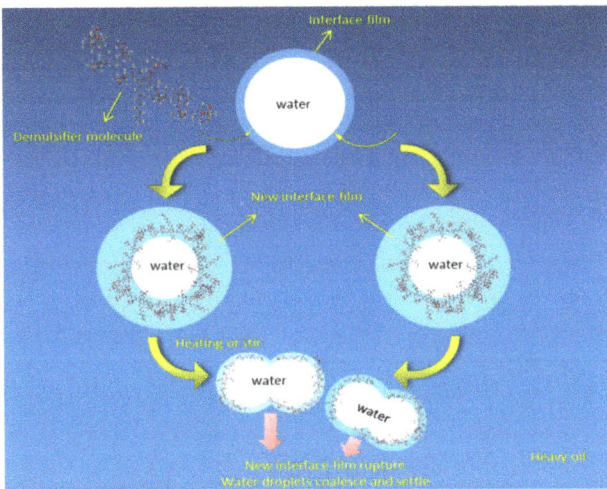

Figure 22. Demulsification mechanism of demulsifier.

4. Conclusions

In this paper, based on the physical parameters of Liaohe crude oil emulsion, 24 kinds of demulsifiers were screened by using the interface generation energy (IFE) module in the molecular dynamics simulation software Materials Studio, and neural network analysis (NNA) and genetic function approximation (GFA) were applied to predict demulsification. The simulation results show that the SDJ9927 demulsifier 6# had the largest reduction in the total energy of the oil–water interface and the strongest reduction in oil–water interfacial tension, and the interfacial formation energy reached −640.48 Kcal/mol. NNA predicted that the water removal amount of the SDJ9927 demulsifier was 7.21 mL, with an overall error of less than 1.83. GFA predicted that the water removal amount of the SDJ9927 demulsifier was 7.41 mL, with an overall error of less than 0.9. The predicted results are consistent with the experimental screening results. SDJ9927 had the highest water removal amount and the best demulsification effect. NNA and GFA had high correlation coefficients, and their R^2s were 0.802 and 0.861, respectively. The higher R^2 was, the more accurate the prediction accuracy was. Finally, the demulsification mechanism of the fluorinated polyether demulsifier was the following: The demulsifier molecules with high interfacial activity replace the natural emulsifier on the oil–water interfacial film and form a new unstable interfacial film. When subjected to heating or shaking, the interfacial film collides with other macromolecules, and the interfacial film breaks, and the water droplets gather to complete the oil–water separation. The demulsification mechanism of the interfacial film was broken by the collision of the fluorinated polyether demulsifier. It was found that when subjected to heating or shaking, the macromolecules in the emulsion exhibited irregular Brownian motion and collided with other macromolecules, resulting in the rupture of the interfacial film. The water in the internal phase broke through the interfacial film and entered the external phase to aggregate, so as to achieve the purpose of oil–water separation.

Author Contributions: Conceptualization, L.W.; methodology, C.L. and L.W.; software, X.G. and H.G.; validation, X.J.; formal analysis, X.G.; investigation, C.S.; resources, Y.C.; data curation, L.H.; writing—original draft, X.G.; writing—review and editing, L.Z. and X.J. All authors have read and agreed to the published version of the manuscript.

Funding: This research was funded by the Natural Science Foundation of Shandong Province for Youth, grant number ZR2020QE111, key projects of the Shandong Natural Science Fund, grant number ZR2020KE041 and Postdoctoral Science Foundation of China, grant number 2020M681073.

Institutional Review Board Statement: Not applicable for studies not involving humans or animals.

Informed Consent Statement: Not applicable.

Data Availability Statement: All data, models, and code generated or used during the study appear in the submitted article.

Acknowledgments: The authors are grateful for the reviewers' instructive suggestions and careful proofreading.

Conflicts of Interest: The authors declare no conflict of interest.

Sample Availability: Samples of the compounds are available from the authors.

References

1. Zhu, G.; Zhang, S.; Liu, Q.; Zhang, J.; Yang, J.; Wu, T.; Huang, Y.; Meng, S. Distribution and treatment of harmful gas from heavy oil production in the Liaohe Oilfield, Northeast China. *Pet. Sci.* **2010**, *7*, 422–427. [CrossRef]
2. Zhang, B.; Zhang, R.; Huang, D.; Shen, Y.; Gao, X.; Shi, W. Membrane fouling in microfiltration of alkali/surfactant/polymer flooding oilfield wastewater: Effect of interactions of key foulants. *J. Colloid Interface Sci.* **2020**, *570*, 20–30. [CrossRef] [PubMed]
3. Ren, L.; Zhang, D.; Chen, Z.; Meng, X.; Gu, W.; Chen, M. Study on Basic Properties of Alkali-Surfactant-Polymer Flooding Water and Influence of Oil-Displacing Agent on Oil–Water Settlement. *Arab. J. Sci. Eng.* **2021**, *27*, 1–9. [CrossRef]
4. Sun, L.; Wu, X.; Zhou, W.; Li, X.; Han, P. Technologies of enhancing oil recovery by chemical flooding in Daqing Oilfield, NE China. *Pet. Explor. Dev.* **2018**, *45*, 673–684. [CrossRef]

5. Feitosa, F.X.; Alves, R.S.; De Sant'Ana, H.B. Synthesis and application of additives based on cardanol as demulsifier for water-in-oil emulsions. *Fuel* **2019**, *245*, 21–28. [CrossRef]
6. Novriansyah, A.; Bae, W.; Park, C.; Permadi, A.K.; Riswati, S.S. Optimal design of alkaline–surfactant–polymer flooding under low salinity environment. *Polymers* **2020**, *12*, 626. [CrossRef]
7. Yang, Y.; Feng, J.; Cao, X.-L.; Xu, Z.-C.; Gong, Q.-T.; Zhang, L.; Zhang, L. Effect of Demulsifier Structures on the Interfacial Dilational Properties of Oil–Water Films. *J. Dispers. Sci. Technol.* **2015**, *37*, 1050–1058. [CrossRef]
8. Sun, H.; Wang, Q.; Li, X.; He, X. Novel polyether-polyquaternium copolymer as an effective reverse demulsifier for O/W emulsions: Demulsification performance and mechanism. *Fuel* **2020**, *263*, 116770. [CrossRef]
9. Yi, M.; Huang, J.; Wang, L. Research on Crude Oil Demulsification Using the Combined Method of Ultrasound and Chemical Demulsifier. *J. Chem.* **2017**, *2017*, 9147926. [CrossRef]
10. Pradilla, D.; Ramírez, J.; Zanetti, F.; Álvarez, O. Demulsifier Performance and Dehydration Mechanisms in Colombian Heavy Crude Oil Emulsions. *Energy Fuels* **2017**, *31*, 10369–10377. [CrossRef]
11. Hjartnes, T.N.; Mhatre, S.; Gao, B.; Sørland, G.H.; Simon, S.; Sjöblom, J. Demulsification of crude oil emulsions tracked by pulsed field gradient NMR. part I: Chemical demulsification. *Ind. Eng. Chem. Res.* **2019**, *58*, 2310–2323. [CrossRef]
12. Wang, D.; Yang, D.; Huang, C.; Huang, Y.; Yang, D.; Zhang, H.; Liu, Q.; Tang, T.; El-Din, M.G.; Kemppi, T.; et al. Stabilization mechanism and chemical demulsification of water-in-oil and oil-in-water emulsions in petroleum industry: A review. *Fuel* **2021**, *286*, 119390. [CrossRef]
13. Shehzad, F.; Hussein, I.A.; Kamal, M.S.; Ahmad, W.; Sultan, A.S.; Nasser, M.S. Polymeric Surfactants and Emerging Alternatives used in the Demulsification of Produced Water: A Review. *Polym. Rev.* **2018**, *58*, 63–101. [CrossRef]
14. Spinelli, L.S.; Aquino, A.S.; Pires, R.V.; Barboza, E.M.; Louvisse, A.M.T.; Lucas, E.F. Influence of polymer bases on the synergistic effects obtained from mixtures of additives in the petroleum industry: Performance and residue formation. *J. Pet. Sci. Eng.* **2007**, *58*, 111–118. [CrossRef]
15. Ahmadi, M.; Chen, Z. Insight into the Interfacial Behavior of Surfactants and Asphaltenes: Molecular Dynamics Simulation Study. *Energy Fuels* **2020**, *34*, 13536–13551. [CrossRef]
16. Gunsteren, W.F.V.; Berendsen, H.J.C. Computer-simulation of molecular-dynamics-methodology, applications, and perspectives in chemistry. *Angew. Chem. Int. Ed.* **2010**, *29*, 992–1023. [CrossRef]
17. Marquez, R.; Forgiarini, A.M.; Langevin, D.; Salager, J.-L. Breaking of Water-In-Crude Oil Emulsions. Part 9. New Interfacial Rheology Characteristics Measured Using a Spinning Drop Rheometer at Optimum Formulation. *Energy Fuels* **2019**, *33*, 8151–8164. [CrossRef]
18. Ballal, D.; Srivastava, R. Modeling the interfacial properties of Poly(Ethylene oxide-Co-Propylene oxide) polymers at water-toluene interface. *Fluid Phase Equilibria* **2016**, *427*, 209–218. [CrossRef]
19. Zhang, L.; Ying, H.; Yan, S.; Zhan, N.; Guo, Y.; Fang, W. Hyperbranched poly(amido amine) as an effective demulsifier for oil-in-water emulsions of microdroplets. *Fuel* **2018**, *211*, 197–205. [CrossRef]
20. Wei, L.; Zhang, L.; Chao, M.; Jia, X.; Liu, C.; Shi, L. Synthesis and Study of a New Type of Nonanionic Demulsifier for Chemical Flooding Emulsion Demulsification. *ACS Omega* **2021**, *6*, 17709–17719. [CrossRef]
21. Ji, S.; Xu, W.; Yang, M.; Yu, K. 3D Convolutional Neural Networks for Human Action Recognition. *IEEE Trans. Pattern Anal. Mach. Intell.* **2013**, *35*, 221–231. [CrossRef] [PubMed]
22. Smyser, C.D.; Inder, T.E.; Shimony, J.S.; Hill, J.E.; Degnan, A.; Snyder, A.Z.; Neil, J.J. Longitudinal Analysis of Neural Network Development in Preterm Infants. *Cereb. Cortex* **2010**, *20*, 2852–2862. [CrossRef] [PubMed]
23. Pramanik, S.; Roy, K. Exploring QSTR modeling and toxicophore mapping for identification of important molecular features contributing to the chemical toxicity in Escherichia coli. *Toxicol. Vitr.* **2014**, *28*, 265–272. [CrossRef]
24. Al-Hajri, M.T.; Abido, M.A.; Darwish, M.K. Assessment of genetic algorithm selection, crossover and mutation techniques in power loss optimization for a hydrocarbon facility. In Proceedings of the 2015 50th International Universities Power Engineering Conference (UPEC), Trent, UK, 1–4 September 2015; pp. 1–6.
25. Sastry, K.; Goldberg, D.E.; Kendall, G. *Genetic Algorithms*; Springer: Boston, MA, USA, 2006.
26. Hanif, N.M.; Adnan, S.N.N.; Latif, M.T.; Zakaria, Z.; Othman, M.R. The composition of surfactants in river water and its influence to the amount of surfactants in drinking water. *World Appl. Sci. J.* **2012**, *17*, 970–975.
27. Wang, Z.-Y.; Gang, H.-Z.; He, X.-L.; Bao, X.-N.; Ye, R.-Q.; Yang, S.-Z.; Li, Y.-C.; Mu, B.-Z. The middle phenyl-group at the hydrophobic tails of bio-based zwitterionic surfactants induced waved monolayers and more hydrated status on the surface of water. *Colloids Surf. A Physicochem. Eng. Asp.* **2021**, *622*, 126655. [CrossRef]
28. Sleiman, S.; Huot, J. Effect of particle size, pressure and temperature on the activation process of hydrogen absorption in TiVZrHfNb high entropy alloy. *J. Alloys Compd.* **2021**, *861*, 158615. [CrossRef]
29. Zhao, Y.; Hu, J.; Li, H.; Zhong, R.; Lin, M. Effect of heavy oil constituents on the oil-water interfacial properties and emulsion stability. *J. Petrochem. Univ.* **2016**, *29*, 32–38.
30. Ma, L.; Zhang, C.; Luo, J. Investigation of the film formation mechanism of oil-in-water (O/W) emulsions. *Soft Matter* **2011**, *7*, 4207–4213. [CrossRef]
31. Frömel, T.; Knepper, T.P. Biodegradation of fluorinated alkyl substances. *Rev. Environ. Contam. Toxicol.* **2010**, *208*, 161–177. [PubMed]
32. Reiner, J.L.; Blaine, A.C.; Higgins, C.P.; Huset, C.; Jenkins, T.M.; Kwadijk, C.J.A.F.; Lange, C.; Muir, D.; Reagen, W.K.; Rich, C.; et al. Polyfluorinated substances in abiotic standard reference materials. *Anal. Bioanal. Chem.* **2015**, *407*, 2975–2983. [CrossRef] [PubMed]

Article

DPD Simulation on the Transformation and Stability of O/W and W/O Microemulsions

Menghua Li [1], Haixia Zhang [1,*], Zongxu Wu [2], Zhenxing Zhu [3] and Xinlei Jia [1,*]

1. Department of Chemical Engineering and Safety, Binzhou University, Binzhou 256603, China; k_mengmeng@163.com
2. Binzhou Dayou New Energy Development Company Limited, Binzhou 256600, China; chnchemwzx@163.com
3. Binzhou City Building and Design Institute, Binzhou 256600, China; zzx1025286003@126.com
* Correspondence: angle00521@163.com (H.Z.); 18436342466@163.com (X.J.)

Abstract: The dissipative particle dynamics simulation method is adopted to investigate the microemulsion systems prepared with surfactant (H1T1), oil (O) and water (W), which are expressed by coarse-grained models. Two topologies of O/W and W/O microemulsions are simulated with various oil and water ratios. Inverse W/O microemulsion transform to O/W microemulsion by decreasing the ratio of oil-water from 3:1 to 1:3. The stability of O/W and W/O microemulsion is controlled by shear rate, inorganic salt and the temperature, and the corresponding results are analyzed by the translucent three-dimensional structure, the mean interfacial tension and end-to-end distance of H1T1. The results show that W/O microemulsion is more stable than O/W microemulsion to resist higher inorganic salt concentration, shear rate and temperature. This investigation provides a powerful tool to predict the structure and the stability of various microemulsion systems, which is of great importance to developing new multifunctional microemulsions for multiple applications.

Keywords: microemulsion; interfacial tension; end to end distance; dissipative particle dynamics (DPD) simulation

1. Introduction

Microemulsion (ME) is a single optically isotropic and thermodynamically stable liquid solution with particle sizes of less than 100 nm and up to 200 nm, consisting of two immiscible liquids such as oil, water or an organic solvent [1]. One phase called the droplet or dispersed phase is embedded in another phase named the continuous phase [2]. According to the spread and continuous phase, microemulsion may be classified into the following three types: water in oil (W/O) that oil is the continuous phase, oil in water (O/W) that water is the continuous phase, and bicontinuous phase, as defined by Liu et al. and Yew et al. [3,4]. Microemulsions with the particular structure are of great importance in many industrial fields, such as the adsorption and the skin penetration of drugs [5–9], oil recovery [10,11], heck reactions [12], and luminescent solar concentrator [13] because of the particular characteristics of ultralow interfacial tension, sizeable interfacial area, high solubilization and low viscosity.

Surfactant is an essential component always used in microemulsion by their amphiphilic nature. A surfactant consisting of a hydrophobic hydrocarbon tail and a hydrophilic polar head group could decrease the interfacial tension and negatively interfere with the phase-separation process to obtain long-term stable emulsion [14–16]. Emulsions are thermodynamic unstable and prone to coalescence, sedimentation, flocculation, and other phenomena. The surfactant molecules exert their role as interface stabilizers, which could migrate toward the oil-water interface and inhibits coalescence [17]. In which, the part of surfactant tail length involves van der Waals interactions between their hydrocarbon chains. Posocco et al. observe the intermolecular hydrophobic interaction forces that are more robust with longer hydrophobic tails (at least eight carbon atoms), eventually, the corresponding interfacial film is more stable [2].

Previous studies about the affecting factors for emulsions including the composition of the emulsion, pH, ionic strength, shear and cryoprotectants on the stability have been carried out over the last years [17–22]. Okuro et al. observed the phase inversion of W/O high internal phase emulsions (HIPEs) to O/W emulsions with higher energy input. Moreover, the shear-thinning behavior and instability were obtained in all W/O-HIPEs at a high shear rate, whereas O/W emulsions showed greater viscosity and stability [23]. Huang et al. observed that the temperature, water–oil ratio and HLB value could influence the emulsion stability and emulsion form as O/W or inverse W/O. The inverse W/O emulsion was found to be the most stable with different affecting factors [24]. Zhong et al. showed that the addition of salt ions resulted in an increased extent of interfacial-protein adsorption and proved to be more durable. When 100 mM salt was added, the emulsions had the best stability [20]. Furthermore, the stability of fuel microemulsions has been investigated by Olsson et al. [25,26], Dash et al. [27,28] and Piskunov et al. [29,30]. These investigations mainly focus on experimental analysis, which is not the mechanism that affects the factors of emulsion formation and stability. Numerical simulations provide a viable strategy to investigate the mechanism in order to overcome practical limitations.

Computer simulations emerged as a powerful tool for studying the microstructures of amphiphilic copolymers [31]. Molecular dynamic simulation has been used more for studying the phase behavior [32–34]. Ma et al. adopted the molecular dynamic simulation to reveal the molecular mechanisms on the stability and instability of the interfacially active asphaltenes (IAA) stabilized O/W emulsions [30]. Dissipative particle dynamics (DPD) is frequently adopted for liquid systems and can reflect the dynamics on the mesoscopic molecular level of complex fluid systems, as one method of molecular dynamics simulation [35–39]. Furthermore, DPD can provide both the equilibrium thermodynamic properties and the dynamic details and the structural changes over time, which are either difficult or impossible to obtain by measurements [40–42]. Rekvig et al. successfully used a DPD simulation to investigate the influence of surfactant branching on the interfacial properties [43]. Wang et al. modified the DPD method, which is an excellent alternative to observe the interfacial properties of surfactant, oil and water systems at various temperatures and salts [44]. Accordingly, herein, the purpose of this study is to reveal the phase inversion and the stability factors of W/O and O/W emulsions using the DPD simulation method. We hope, according to this work, to observe a new way to form a more stable microemulsion and broaden the applications of microemulsions in many industries.

2. Simulation Methodology
2.1. Dissipative Particle Dynamics Theory

Dissipative particle dynamics (DPD) was firstly reported by Koelman and Hoogerbrugge as an efficient mesoscopic-level simulation method [36]. Several atoms or molecules are represented by beads that interact with each other via effective pair potentials. To simplify the calculations, the beads have the same mass, length, and time scales, in which the mass of the beads equals to 1 DPD unit. Every two beads i and j in a system interact with each other by the following formula from Groot [37]:

$$f_{ij} = F_{ij}^{C}(r_{ij}) + F_{ij}^{R}(r_{ij}) + F_{ij}^{D}(r_{ij}) \tag{1}$$

where F_{ij}^{C}, F_{ij}^{R} and F_{ij}^{D} denote a conservative force, a random force and a dissipative force, respectively. In which, F_{ij}^{C} contains a harmonic spring force (F_{ij}^{Cr}) and a soft repulsion force (F_{ij}^{Cs}), which are given by

$$\begin{aligned} F_{ij}^{Cr} &= \alpha_{ij}\left(1 - \frac{r_{ij}}{r_c}\right)\frac{\hat{r}_{ij}}{r_{ij}}(r_{ij} < r_c) \\ &= 0(r_{ij} > r_c) \end{aligned} \tag{2}$$

and
$$F_{ij}^{Cs} = -C \cdot \hat{r}_{ij} \tag{3}$$

Among them, α_{ij} is the maximum repulsion parameters between particles i and j, $\hat{r}_{ij} = \hat{r}_i - \hat{r}_j$, $r_{ij} = |\hat{r}_{ij}|$, r_{ij} is the distance between i and j, with the corresponding unit vector $\hat{r}_{ij}$, r_c is a cutoff radius which provides the extent of the interaction range and C is the spring constant. Moreover, the random force (F_{ij}^R) and the dissipative force (F_{ij}^D) can be shown by the following equations from Groot [37]:

$$F_{ij}^R = \sigma \omega^R(r_{ij}) \theta_{ij} \hat{r}_{ij} \tag{4}$$

$$F_{ij}^D = -\eta \omega^D(r_{ij})(r_{ij} \cdot v_{ij}) \theta_{ij} \hat{r}_{ij} \tag{5}$$

Here, θ_{ij} is the random fluctuation variable between 0 and 1, v_{ij} represents the relative velocities of the beads, and ω is the weight function. Furthermore, h is the friction coefficient and s is the noise amplitude, and $\sigma^2 = 2\eta k_B T$. To sample the canonical ensemble distribution, s, h and α_{ij} determine the amplitude of the dissipative, conservative and random forces [44].

$\omega^D = (\omega^R)^2$ was made to comply with the fluctuation-dissipation theorem, and the temperature follows from the relation between h and s. The same parameters, weight functions, and integration algorithm were used from Groot and Warren [37]:

$$\omega^C(r_{ij}) = \omega^R(r_{ij}) = \sqrt{\omega^D(r_{ij})} = \omega(r_{ij}) \tag{6}$$

where
$$\omega(r_{ij}) = \begin{cases} 1 - \frac{r}{R_C}(r < R_C) \\ 0(r \geq R_C) \end{cases} \tag{7}$$

A modified version of the velocity verlet algorithm is adopted in the Newton's equations of motion and the reduced units are used in our paper. Cutoff radius R_c, $k_B T$ and m of the particles are used as the unit of length, energy and mass, respectively. Here, $k_B T$ represents the micro temperature, in which k_B is boltzmann constant and T the thermodynamic temperature. h = 4.5 and s = 3 are set in our research.

2.2. Models and Interaction Parameters

Water, oil of n-hexane and surfactant of sodium lauryl sulfate (SDS) were included in our research system. The coarse-grained models and shorthand notation for each molecular are presented in Figure 1. SDS are separated into two groups of hydrophilic and hydrophobic parts with the beads H set in green and T designated in blue, respectively. The H and T connected by a harmonic spring can be denoted by the symbol H1T1. Water is bead W in red, n-hexane is bead O in rose.

To clearly show the simulation results, the periodic boundary condition of three directions was employed in the cubic simulation box, which was 15 × 15 × 15 R_c^3 (L_x × L_y × L_z). There were approximately 10,125 beads in every simulation box, and the density of beads was set to ρ = 3.0. It is possible to convert the simulation surfactant concentration to the mole fraction with the isochoric property. Based on Groot's reports [37], the spring constant of every bead was set to 4.0.

The diffusivities of beads changed with an increasing simulation time, and are shown in Figure 2. A gradual decrease was observed with an increasing simulation time until 600 DPD units, and then remained unchanged after 800 DPD units. The simple modification was conducted following the velocity–varlet algorithm reported by Groot and Warren [37] and set Δt = 0.05. Therefore, it is an equilibrium state can be reached by 20,000 timesteps simulations.

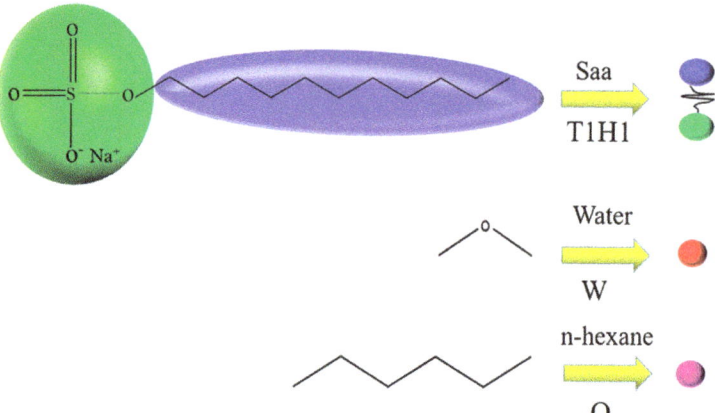

Figure 1. A coarse-grained model for surfactant, water and oil.

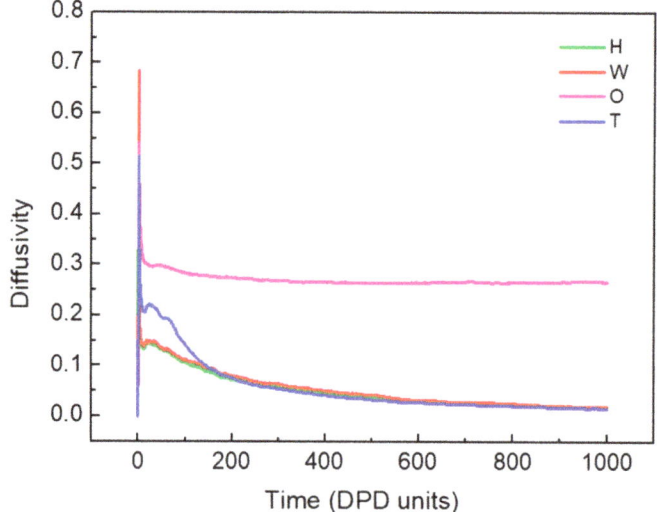

Figure 2. The diffusivity of beads H, T, W and O with the simulation time. The oil-water ratio was equal to 3:1.

Table 1 is the repulsive interaction parameters between different beads referred to the previous reports [40,41,45].

Table 1. The interaction parameters employed in this simulation.

	W	H	T	O
W	25			
H	25.34	25		
T	151.52	177.82	25	
O	103.24	143.61	25.94	25

3. Results and Discussion

3.1. Transformation of W/O and O/W Microemulsion Systems

3.1.1. Dynamics of the W/O and O/W Microemulsion Systems Formation

A rigorous strategy for the formation of the W/O and O/W microemulsion systems is proposed, which provides the microstructure of formed microemulsions. We chose pure water and n-hexane as two incompatible systems, sodium lauryl sulfate as a surfactant to simulate the microemulsion by dissipative particle dynamics (DPD) simulation. The typical snapshots illustrating the evolution of the microemulsion structure with time as an example are presented in Figure 3, where the surfactant concentration is 0.1 and the value of oil-water is 1/3. To better observe the internal structure of the microemulsion, the oil beads were not exhibited in the simulation system. In the initial state (a), surfactants, oil and water were randomly dispersed in the simulation box. At a more extensive simulation time of 250 DPD units (b), the monolayer becomes gathered and the surfactant hydrophilic group associate with the water molecules to form a sheet aggregate. A further increase in the simulation time of 500 DPD units leads to the formation of the W/O microemulsion (c). Most surfactant molecules associate in the oil-water interfaces compactly and little surfactant molecules associate to form a little micelle in the system. Surprisingly, the little micelle disappears with increasing the simulation time to 800 DPD units (d), indicating that the interface of the microemulsion approaches the maximum interface concentration and forms a stable microemulsion. Therefore, the simulation time of 1000 DPD units is enough to create the microemulsion.

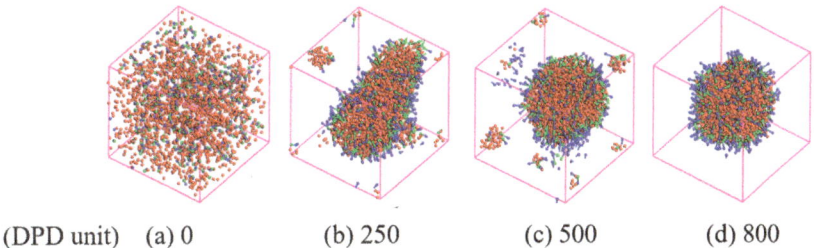

(DPD unit)   (a) 0   (b) 250   (c) 500   (d) 800

Figure 3. Snapshots of the evolution of the W/O microemulsion structure in DPD simulations with oil/water = 1/3. The hydrophilic group of surfactants are shown in green; the hydrophobic group is shown in blue. The water beads is shown in red. To clarity observe the internal structure, oil beads (pink) are not displayed.

3.1.2. Influence of Oil-Water Ratio on the Transformation of W/O and O/W Microemulsion Systems

The Oil-Water ratio could influence the emulsion stability and the emulsion formation of O/W or inverse W/O under high-energy input, which has extra stability with various affecting factors [23]. We set up the value of oil/water from 4:1 to 1:5 to study the microemulsion type affected by the oil/water ratio with the surfactant concentration of 0.05. Figure 4 shows the translucent three-dimensional structure of the simulation system (a), the corresponding mean interfacial tension (b) and end to end distance of H1T1 (c). The Irving and Kirkwood (IK) method was adopted to analyze the mean interfacial tension by:

$$r_{sim} = \frac{1}{2} \int_{-L_Z/2}^{L_Z/2} [P_N(Z) - P_L(Z)]dz \quad (8)$$

Among them, $P_N(Z)$ was the pressure normal to the interface, the same as $P_{zz}(Z)$. The lateral force was given by $P_L(Z) = 1/2[P_{xx}(Z) + P_{yy}(Z)]$ with the pressure tensor component in the Z direction. The reality units could be transformed from the mean surface tension satisfied using the simulations by $\gamma = \gamma_{sim} \times k_B T/R_c^2$, with R_c = 0.711 nm and T = 298 K [39].

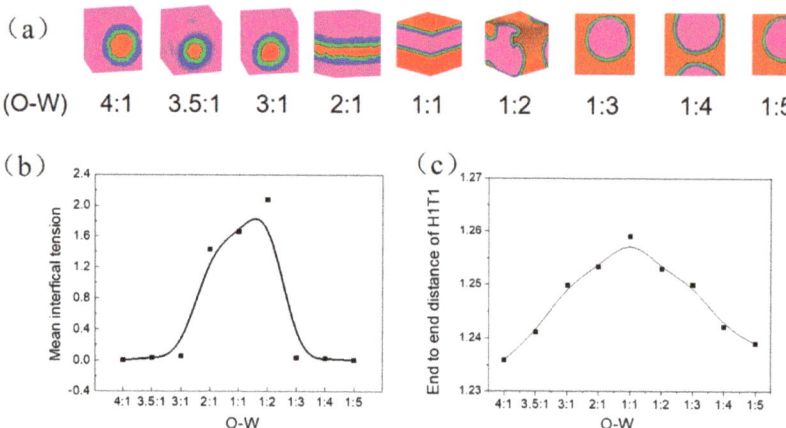

Figure 4. The translucent three-dimensional structure of the simulation system (**a**), the corresponding mean interfacial tension (**b**) and end to end distance of H1T1 (**c**).

As the same content of oil and water as the oil/water = 1/1, surfactants adsorb at a flat interface and form a layer-like aggregate. The highest mean interfacial tension and end to end distance of H1T1 indicate that the surfactant molecular chain is most extended and has the weakest surfactant activity and emulsification capacity. With increasing oil content to the oil/water = 2/1, the increased oil phase with small amount was not sufficient to change the layer oil-water interface. The oil phase is sufficient to wrap the water phase and begin to form W/O microemulsion when the oil content increases to the oil/water = 3/1. Simultaneously, the decrease of the mean interfacial tension and end to end distance of H1T1 indicate the surfactant molecular chain owns a degree of bending and better emulsifying capacity. It is mainly due to the reduced oil and water interface with a larger oil phase and smaller water phase at the same surfactant concentration. As the oil phase increased to 3.5/1 and 4/1, the W/O microemulsion proved to be more stable, with a shrinking end to end distance of H1T1; however, the mean interfacial tension remained unchanged. Instead, O/W microemulsion was formed by increasing the water content to oil/water = 1:3. The mean interfacial tension reached its minimum value and did not change despite the addition of more water molecules. A gradual decrease in the end to end distance of H1T1 shows that the surfactant molecules were more compact and orderly at the interface and the O/W microemulsion is more stable.

Figure 5 shows the density distributions of beads (H, T, W, O) along the x-axis in different oil-water ratios (3/1 and 1/3) corresponding the translucent three-dimensional structure used to observe the actual adsorption of surfactant molecules at the oil and water interface. W/O microemulsion is formed with the raised curve of W shown in Figure 5a, and the droplet size is around 12 DPD units (from 2 to 14 DPD units). Meanwhile, the internal water phase associated with the hydrophilic head makes the density of H slightly higher than T, which is caused by the smaller space of the internal water phase than the outside oil phase, with the same amounts of H and T. The translucent three-dimensional structure (Figure 5b) could observe this structure more intuitively. In contrast, the raised curve of O and T was slightly higher than H (Figure 5c) indicate the formation of O/W microemulsion. Figure 5d shows the inner structure of the formed O/W microemulsion, of which the hydrophobic group associated with the surface of the oil phase and the hydrophilic group disperse in the water phase.

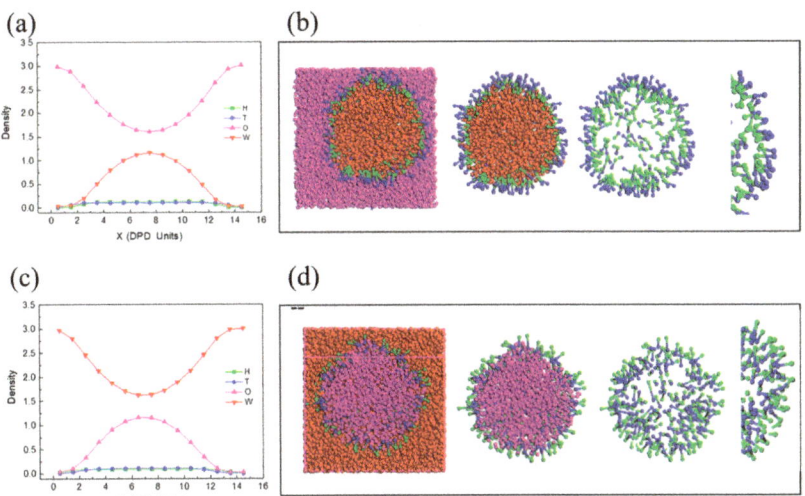

Figure 5. The density distribution of beads (H, T, W, O) along the x-axis in different oil-water ratios corresponding the translucent three-dimensional structures: (**a**,**b**) O/W = 3/1, (**c**,**d**) O/W = 1/3.

3.2. Stability of the W/O and O/W Microemulsion Systems

3.2.1. Influence of Temperature on the W/O and O/W Microemulsion Systems

For safety, some food products based on emulsion often require heating treatment such as cooking and pasteurization. However, the emulsion could transform from W/O emulsion to O/W emulsion by changing the temperature from that observed in previous studies [18,24,44]. Therefore, the stability of the emulsion will be affected by the temperature. Figure 6 shows the impact of temperature on the W/O and O/W microemulsions. From the translucent three-dimensional structure of O/W microemulsion (Figure 6a), we can observe the minimum value for the stability of O/W microemulsion at 0.8 k_BT, but the maximum value at 1.0 k_BT with the minimal mean interfacial tension (Figure 6c). However, a greater increase is observed in the mean interfacial tension after heating over 1.0 k_BT or cooling to 0.8 k_BT. This may be due to the O/W microemulsion transformed into a rod structure with a more extensive oil-water interface than microemulsion, which reduces the ability to change the surface activity of H1T1. End to end distance of H1T1 (Figure 6e) gradually increases with increasing the temperature caused by more extension molecular chain consistent with the results obtained by Chen et al. [46].

By contrast, the W/O microemulsion demonstrated a better stability resistance to temperature as long as the temperature is below 1.55 k_BT, as shown in Figure 6b. However, one big W/O droplet was separated into a number of tiny droplets when the temperature decreased to 0.1 k_BT. Compared with O/W microemulsion, the better stability of W/O microemulsion may be due to the higher viscosity of oil molecules that were not easily spread in the decentralized system.

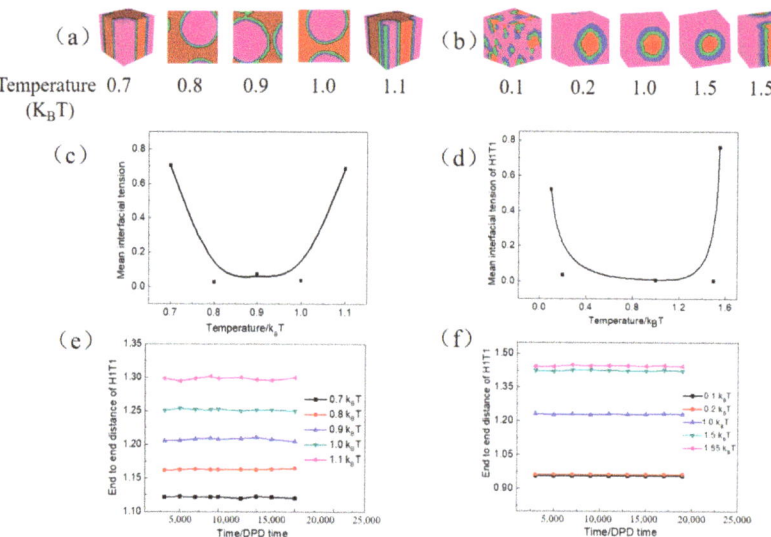

Figure 6. The translucent three-dimensional structure of microemulsion—(**a**,**b**), the corresponding mean interfacial tension—(**c**,**d**) and end to end distance of H1T1—(**e**,**f**) with increasing temperature. In which, (**a**,**c**,**e**) represent O/W microemulsion, (**b**,**d**,**f**) represent W/O microemulsion.

3.2.2. Influence of Inorganic Salt on the W/O and O/W Microemulsion Systems

Two salts (NaCl and CaCl$_2$) affect the emulsion stability and were observed using experiment and simulation by Zhong et al. [20] and Zhang et al. [47]. Despite this, we investigated the influence of salt on the stability of W/O and O/W microemulsion by the DPD simulation shown in Figure 7. The decrease in head–head repulsion parameters (α_{HH}) means adding the inorganic salt, and if no inorganic salt exists α_{HH} = 25. The translucent three-dimensional structure of O/W microemulsion (Figure 7a) exhibits that the droplet transform into the rod topology with decreasing α_{HH} to 22; meanwhile, a significant increase occurred in the mean interfacial tension (Figure 7c). There are certain fluctuations in end to end distance of H1T1 with increasing simulation time whether the inorganic salt is added shown in Figure 7e. This is mainly due to the different degree of adsorption of surfactant molecules at the oil-water interface. At the beginning of the simulation, the surfactant molecules were scattered in an aqueous solution and showed a certain degree of bending because of the intermolecular repulsion. With the increase in simulation time, small unstable microemulsion droplets gradually formed and finally reached the equilibrium state to form one stable and large microemulsion droplet and resulted in different degrees of bending. However, end to end distance of H1T1 gradually increases with increasing salt concentration (Figure 7e). It was probably caused by the opposite ion of inorganic salt could neutralize part of the charge of the hydrophilic group and reduce the electrostatic repulsion between the hydrophilic groups. It leads to looser surfactant molecules at the oil and water interface and the molecular chains are more extended.

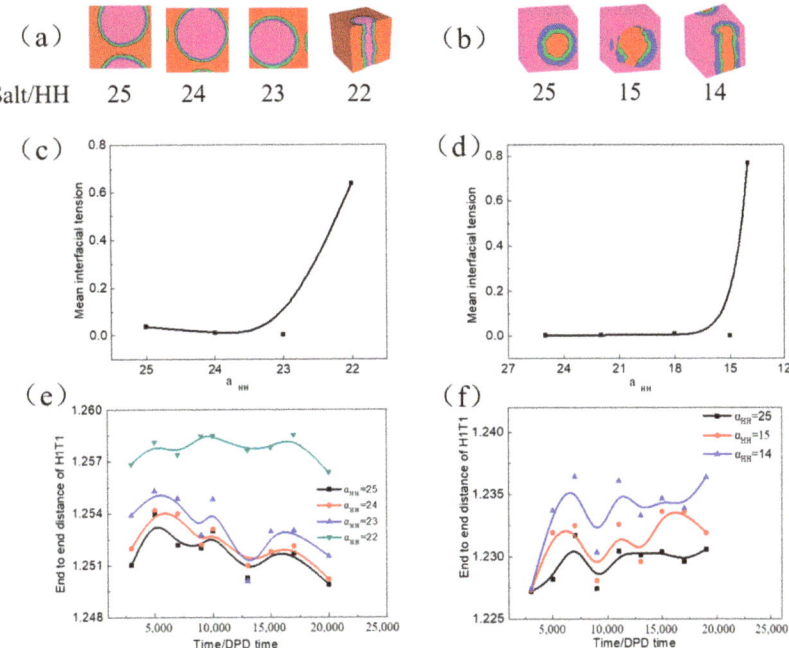

Figure 7. The translucent three-dimensional structure of microemulsion—(**a**,**b**), the corresponding mean interfacial tension—(**c**,**d**) and end to end distance of H1T1—(**e**,**f**) with increasing inorganic salt. In which, (**a**,**c**,**e**) represent O/W microemulsion, (**b**,**d**,**f**) represent W/O microemulsion.

However, W/O microemulsion begins to transform when α_{HH} decreases to 14 (Figure 7b). The rod topology has the highest interface tension (Figure 7d) and most significant end-to-end distance (Figure 7f). W/O microemulsion has a better stability resistance to the inorganic salt compared with O/W microemulsion, which is consistent with the results observed by Huang et al. [24]. This may be due to the adsorption position of H1T1 at the oil and water interfaces. The hydrophilic group associated with the interface caused a more significant spatial block effect and hindered the electrostatic gravity of inorganic salt and hydrophilic ions in W/O microemulsion.

3.2.3. Influence of Shear on the W/O and O/W Microemulsion Systems

Lee–Edwards sliding-brick boundary conditions along x-axis were applied to the simulation system representing the impact of shear flow on the structure and orientation of complex fluids. Figure 8 reveals the influence of shear on the O/W and W/O microemulsion. The topology of the O/W microemulsion droplet is stable in the absence of shear and maintains this structure when increasing the shear rate to $0.008 s^{-1}$. However, the oil cannot be wrapped by water after the shear rate is enhanced to $0.009 s^{-1}$ and transformed to a layer-like aggregate (Figure 8a) with a significant increase in the mean interfacial tension (Figure 8c). This is mainly due to the shear rate inducing the aggregate spread out along the direction of the shear rate, similarly to an external force to the microemulsion. An increase in the end to end distance of H1T1 for O/W microemulsion indicates that the molecular chain is more extended with a larger shear rate (Figure 8e). However, for W/O microemulsion, the droplet emerges deformation until the shear rate to $0.031 s^{-1}$ (Figure 8b) with a more considerable mean interfacial tension (Figure 8d) and end to end distance of H1T1 (Figure 8f). In light of this, W/O microemulsion has better stability to resist the shear rate owing to the higher viscosity of the oil phase as the continuous phase is not easy to affect by an external force.

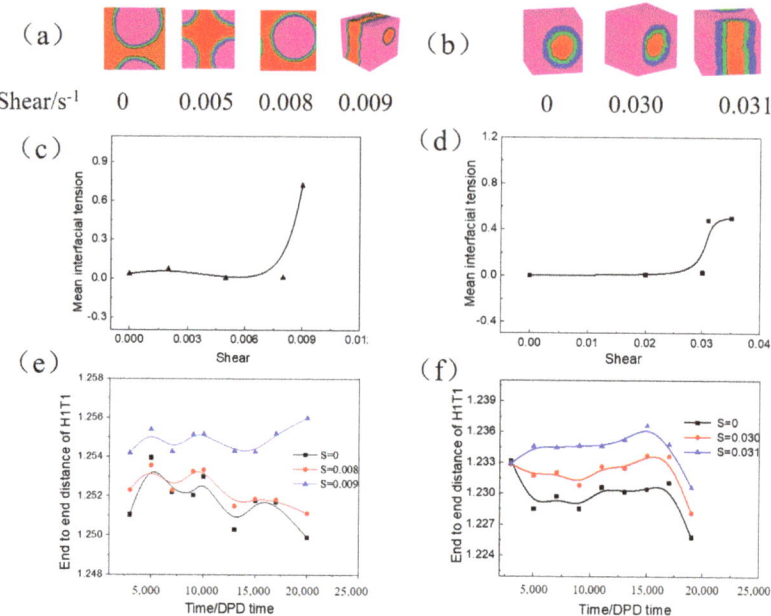

Figure 8. The translucent three-dimensional structure of microemulsion—(**a**,**b**), the corresponding mean interfacial tension—(**c**,**d**) and end to end distance of H1T1—(**e**,**f**) with increasing shear rate. In which, (**a**,**c**,**e**) represent O/W microemulsion, (**b**,**d**,**f**) represent W/O microemulsion.

4. Conclusions

The topology and stability of O/W and inverse W/O microemulsion were studied using the dissipative particle-dynamics simulation method. Coarse-grained models were constructed for surfactant (H1T1), oil and water, respectively. The results show that the ratio of oil and water would change the topology of the microemulsion that transforms from W/O to O/W by decreasing the value of oil/water from 3:1 to 1:3. Meanwhile, the effects of the temperature, inorganic salt and shear rate on the stability of the formed microemulsions were researched with a translucent three-dimensional structure and corresponding parameters such as mean interfacial tension and end-to-end distance of H1T1. Inverse W/O microemulsion has better resistance to a higher temperature (1.5 k_BT), inorganic salt (α_{HH} = 14) and shear rate (0.03 s^{-1}) than the O/W microemulsion of T = 1.0 k_BT, α_{HH} = 23 and s = 0.008 s^{-1}. In the inverse W/O microemulsion of oil as the continuous phase with higher viscosity is not easy to affect the physical and chemical properties. The simulation provides a powerful tool to forecast the structure and the stability of various microemulsions, which is of great importance for developing new functional emulsions for many applications.

Author Contributions: Conceptualization, H.Z. and X.J.; methodology, H.Z.; software, H.Z.; validation, H.Z., Z.Z. and Z.W.; formal analysis, H.Z.; investigation, M.L.; resources, M.L.; data curation, M.L.; writing—original draft preparation, M.L.; writing—review and editing, H.Z.; visualization, Z.Z.; supervision, Z.W.; project administration, H.Z.; funding acquisition, H.Z. All authors have read and agreed to the published version of the manuscript.

Funding: The work was supported by the Research Foundation Project under Grant [BZXYLG2121]; the Key Technology Research and Development Program of Shandong Province under Grant [2019GSF109117].

Institutional Review Board Statement: Not applicable.

Informed Consent Statement: Not applicable.

Conflicts of Interest: The authors declare no conflict of interest.

Sample Availability: Samples of all the compounds are available from the authors.

References

1. Huang, Y.; Ni, N.; Zhao, L.; Zhang, J.; Shen, L. The preparation, evaluation and phase behavior of linker-based coix seed oil microemulsion. *J. Mol. Liq.* **2020**, *319*, 114314. [CrossRef]
2. Posocco, P.; Perazzo, A.; Preziosi, V.; Laurini, E.; Pricl, S.; Guido, S. Interfacial tension of oil/water emulsions with mixed non-ionic surfactants: Comparison between experiments and molecular simulations. *RSC Adv.* **2016**, *6*, 4723–4729. [CrossRef]
3. Liu, J.; Wu, J.; Sun, J.; Wang, D.; Wang, Z. Investigation of the phase behavior of food-grade microemulsions by mesoscopic simulation. *Colloid Surf. A* **2015**, *487*, 75–83. [CrossRef]
4. Yew, H.C.; Misran, M.B. Nonionic Mixed Surfactant Stabilized Water-in-Oil Microemulsions for Active Ingredient In Vitro Sustained Release. *J. Surfactants Deterg.* **2015**, *19*, 49–56. [CrossRef]
5. Subongkot, T.; Ngawhirunpat, T. Development of a novel microemulsion for oral absorption enhancement of all-trans retinoic acid. *Int. J. Nanomed.* **2017**, *12*, 5585–5599. [CrossRef] [PubMed]
6. Subongkot, T.; Sirirak, T. Development and skin penetration pathway evaluation of microemulsions for enhancing the dermal delivery of celecoxib. *Colloids Surf. B Biointerfaces* **2020**, *193*, 111103. [CrossRef]
7. Gundogdu, E.; Alvarez, I.G.; Karasulu, E. Improvement of effect of water-in-oil microemulsion as an oral delivery system for fexofenadine: In vitro and in vivo studies. *Int. J. Nanomed.* **2011**, *6*, 1631–1640. [CrossRef]
8. You, J.; Meng, S.; Ning, Y.K.; Yang, L.Q.; Zhang, X.W.; Wang, H.N.; Li, J.J.; Yin, F.M.; Liu, J.; Zhai, Z.Y.; et al. Development and application of an osthole microemulsion hydrogel for external drug evaluation. *J. Drug Deliv. Sci. Technol.* **2019**, *54*, 101331. [CrossRef]
9. Jaimes-Lizcano, Y.A.; Wang, Q.; Rojas, E.C.; Papadopoulos, K.D. Evaporative destabilization of double emulsions for effective triggering of release. *Colloid Surf. A* **2013**, *423*, 81–88. [CrossRef]
10. Zhu, T.; Kang, W.; Yang, H.; Li, Z.; Zhou, B.; He, Y.; Wang, J.; Aidarova, S.; Sarsenbekuly, B. Advances of microemulsion and its applications for improved oil recovery. *Adv. Colloid Interface Sci.* **2021**, *299*, 102527. [CrossRef]
11. Zhou, Y.; Yin, D.; Wang, D.; Zhang, C.; Yang, Z. Experiment investigation of microemulsion enhanced oil recovery in low permeability reservoir. *J. Mater. Res. Technol.* **2020**, *9*, 8306–8313. [CrossRef]
12. Mangaiyarkarasi, R.; Priyanga, M.; Santhiya, N.; Umadevi, S. In situ preparation of palladium nanoparticles in ionic liquid crystal microemulsion and their application in Heck reaction. *J. Mol. Liq.* **2020**, *310*, 113241. [CrossRef]
13. Congiu, A.; Gila, L.; Caccianotti, L.; Fusco, R.; Busto, C.; Zanardi, S.; Salvalaggio, M. Microemulsions for luminescent solar concentrator application. *Sol. Energy* **2019**, *216*, 338–350. [CrossRef]
14. Fatma, N.; Ansari, W.H.; Panda, M.; Kabir-ud-Din. A Systematic Study of Mixed Surfactant Solutions of a Cationic Ester-Bonded Dimeric Surfactant with Cationic, Anionic and Nonionic Monomeric Surfactants in Aqueous Media. *J. Surfactants Deterg.* **2013**, *16*, 609–620. [CrossRef]
15. Wu, J.; Mei, P.; Chen, W.; Li, Z.B.; Tian, Q.; Mei, Q.X. Surface Properties and Solubility Enhancement of Anionic/Nonionic Surfactant Mixtures Based on Sulfonate Gemini Surfactants. *J. Surfactants Deterg.* **2019**, *22*, 1331–1342. [CrossRef]
16. Khimani, M.; Vora, S. Effect of Inorganic Additives on a Conventional Anionic-Nonionic Mixed Surfactants System in Aqueous Solution. *J. Surfactants Deterg.* **2011**, *14*, 545–554. [CrossRef]
17. Pan, Y.; Xu, Y.; Zhu, L.; Liu, X.; Zhao, G.; Wang, S.; Yang, L.; Ma, T.; Liu, H. Stability and rheological properties of water-in-oil (W/O) emulsions prepared with a soyasaponin-PGPR system. *Future Foods* **2021**, *4*, 100096. [CrossRef]
18. Li, J.Q.; Li, Y.F.; Zhong, J.F.; Teng, J.; Zang, H.; Song, H. Effect of cellulose nanocrystals on the formation and stability of oil in water emulsion formed by octenyl succinic anhydride starch. *LWT* **2021**, *151*, 112214. [CrossRef]
19. Seddari, S.; Moulai-Mostefa, N.; Sabbache, H. Effect of pH on the stability of W/O/W double emulsions prepared by the mixture of biopolymers using direct method. *Mater. Today Proc.* **2021**, *49*, 1030–1034. [CrossRef]
20. Zhong, M.; Sun, Y.; Sun, Y.; Huang, Y.; Qi, B.; Li, Y. The effect of salt ion on the freeze-thaw stability and digestibility of the lipophilic protein-hydroxypropyl methylcellulose emulsion. *LWT* **2021**, *151*, 112202. [CrossRef]
21. Ravera, F.; Dziza, K.; Santini, E.; Cristofolini, L.; Liggieri, L. Emulsification and emulsion stability: The role of the interfacial properties. *Adv. Colloid Interface Sci.* **2021**, *288*, 102344. [CrossRef] [PubMed]
22. Li, L.; He, M.; Yang, H.; Wang, N.; Kong, Y.; Li, Y.; Teng, F. Effect of soybean lipophilic protein–methyl cellulose complex on the stability and digestive properties of water-in-oil-in-water emulsion containing vitamin B12. *Colloid Surf. A* **2021**, *629*, 127104. [CrossRef]
23. Okuro, P.K.; Gomes, A.; Costa, A.L.R.; Adame, M.A.; Cunha, R.A. Formation and stability of W/O-high internal phase emulsions (HIPEs) and derived O/W emulsions stabilized by PGPR and lecithin. *Food Res. Int.* **2019**, *122*, 252–262. [CrossRef]
24. Huang, W.; Zhu, D.; Fan, Y.; Xuea, X.; Yanga, H.; Jiang, L.; Jiang, Q.; Chen, J.; Jiang, B.; Komarneni, S. Preparation of stable inverse emulsions of hydroxyethyl methacrylate and their stability evaluation by centrifugal coefficient. *Colloid Surf. A* **2020**, *604*, 125309. [CrossRef]

25. Kayali, I.; Karaein, M.; Qamhieh, K.; Wadaah, S.; Ahmad, W.; Olsson, U. Phase Behavior of Bicontinuous and Water/Diesel Fuel Microemulsions Using Nonionic Surfactants Combined with Hydrophilic Alcohol Ethoxylates. *J. Dispers. Sci. Technol.* **2015**, *36*, 10–17. [CrossRef]
26. Kayali, I.; Karaein, M.; Ahmed, W.; Qamhieh, K.; Olsson, U. Alternative Diesel Fuel: Microemulsion Phase Behavior and Combustion Properties. *J. Dispers. Sci. Technol.* **2016**, *37*, 894–899. [CrossRef]
27. Acharya, B.; Guru, P.S.; Dash, S. Tween-80-n-Butanol-Diesel-Water Microemulsion System-A Class of Alternative Diesel Fuel. *J. Dispers. Sci. Technol.* **2014**, *35*, 1492–1501. [CrossRef]
28. Acharya, B.; Dash, S. Tuning commercial diesel to microemulsified and blended form: Phase behavior and implications. *J. Dispers. Sci. Technol.* **2019**, *40*, 1159–1168. [CrossRef]
29. Ashikhmin, A.; Piskunov, M.; Yanovsky, V.; Yan, W.M. Properties and Phase Behavior of Water-in-Diesel Microemulsion Fuels Stabilized by Nonionic Surfactants in Combination with Aliphatic Alcohol. *Energy Fuels* **2020**, *34*, 2135–2142. [CrossRef]
30. Ashihmin, A.; Piskunov, M.; Roisman, I.; Yanovsky, V. Thermal stability control of the water-in-diesel microemulsion fuel produced by using a nonionic surfactant combined with aliphatic alcohols. *J. Dispers. Sci. Technol.* **2020**, *41*, 771–778. [CrossRef]
31. Lbadaoui-Darvas, M.; Garberoglio, G.; Karadima, K.S.; Cordeiro, M.N.D.S.; Nenes, A.; Takahama, S. Molecular simulations of interfacial systems: Challenges, applications and future perspectives. *Mol. Simul.* **2021**, 1–38. [CrossRef]
32. Ma, J.; Yang, Y.L.; Li, X.G.; Sun, H.; He, L. Mechanisms on the stability and instability of water in oil emulsion stabilized by interfacially active asphaltenes: Role of hydrogen bonding reconstructing. *Fuel* **2021**, *297*, 120763. [CrossRef]
33. Ma, J.; Song, X.Y.; Peng, B.L.; Zhao, T.; Luo, J.H.; Shi, R.F.; Zhao, S.L.; Liu, H.L. Multiscale molecular dynamics simulation study of polyoxyethylated alcohols self assembly in emulsion systems. *Chem. Eng. Sci.* **2020**, *231*, 116252. [CrossRef]
34. Wang, F.X.; Cao, J.H.; Ling, Z.Y.; Zhang, Z.G.; Fang, X.M. Experimental and simulative investigations on a phase change material nano emulsion based liquid cooling thermal management system for a lithium ion battery pack. *Energy* **2020**, *207*, 118215. [CrossRef]
35. Maiti, A.; McGrother, S. Bead-bead interaction parameters in dissipative particle dynamics: Relation to bead-size, solubility parameter, and surface tension. *J. Chem. Phys.* **2004**, *120*, 1594–1601. [CrossRef] [PubMed]
36. Hoogerbrugge, P.J.; Koelman, J.M.V.A. Simulating Microscopic Hydrodynamic Phenomena with Dissipative Particle Dynamics. *Europhys. Lett.* **1992**, *19*, 155–160. [CrossRef]
37. Groot, R.D.; Madden, T.J. Dynamic simulation of diblock copolymer microphase separation. *J. Chem. Phys.* **1998**, *108*, 8713–8724. [CrossRef]
38. Lavagnini, E.; Cook, J.L.; Warren, P.B.; Williamson, M.J.; Hunter, C.A. A Surface Site Interaction Point Method for Dissipative Particle Dynamics Parametrization: Application to Alkyl Ethoxylate Surfactant Self-Assembly. *J. Phys. Chem. B* **2020**, *124*, 5047–5055. [CrossRef]
39. Wang, X.; Santo, K.P.; Neimark, A.V. Modeling Gas-Liquid Interfaces by Dissipative Particle Dynamics: Adsorption and Surface Tension of Cetyl Trimethyl Ammonium Bromide at the Air-Water Interface. *Langmuir* **2020**, *36*, 14686–14698. [CrossRef]
40. Wang, Y.; Wang, H.B.; Li, C.L.; Sun, S.Q.; Hu, S.Q. CO_2 responsive Pickering emulsion stablized by modified silica nanoparticles A dissipative particle dynamics simulation study. *J. Ind. Eng. Chem.* **2021**, *97*, 492–499. [CrossRef]
41. Zhang, H.; Xu, B.; Zhang, H. Mesoscopic simulation on the microemulsion system stabilized by bola surfactant. *J. Dispers. Sci. Technol.* **2021**. [CrossRef]
42. Zhang, H.; Li, D.; Pei, L.; Zhang, L.J.; Wang, F. The Stability of the Micelle Formed by Chain Branch Surfactants and Polymer Under Salt and Shear Force: Insight from Dissipative Particle Dynamics Simulation. *J. Dispers. Sci. Technol.* **2015**, *37*, 270–279. [CrossRef]
43. Rekvig, L.; Kranenburg, M.; Hafskjold, B.; Smit, B. Effect of surfactant structure on interfacial properties. *Europhys. Lett.* **2003**, *63*, 902–907. [CrossRef]
44. Wang, S.; Yang, S.; Wang, R.; Yan, Z. Dissipative particle dynamics study on the temperature dependent interfacial tension in surfactant-oil-water mixtures. *J. Pet. Sci. Eng.* **2018**, *169*, 81–95. [CrossRef]
45. Zhang, H.; Zhu, Z.; Wu, Z.; Wang, F.; Xu, B.; Wang, S.; Zhang, L. Investigation on the formation and stability of microemulsions with Gemini surfactants: DPD simulation. *J. Dispers. Sci. Technol.* **2021**. [CrossRef]
46. Chen, Z.; Cheng, X.; Cui, H.; Cheng, P.; Wang, H. Dissipative particle dynamics simulation of the phase behavior and microstructure of CTAB/octane/1-butanol/water microemulsion. *Colloid Surf. A* **2007**, *301*, 437–443. [CrossRef]
47. Zhang, Y.; Yang, N.; Xu, Y.; Wang, Q.; Huang, P.; Nishinari, K.; Fang, Y. Improving the Stability of Oil Body Emulsions from Diverse Plant Seeds Using Sodium Alginate. *Molecules* **2019**, *24*, 3856. [CrossRef]

Article
Self-Diffusion in Simple Liquids as a Random Walk Process

Sergey A. Khrapak

Joint Institute for High Temperatures, Russian Academy of Sciences, 125412 Moscow, Russia; sergey.khrapak@gmx.de

Abstract: It is demonstrated that self-diffusion in dense liquids can be considered a random walk process; its characteristic length and time scales are identified. This represents an alternative to the often assumed hopping mechanism of diffusion in the liquid state. The approach is illustrated using the one-component plasma model.

Keywords: self-diffusion in liquids; transport properties of liquids; random walk process; viscosity of liquids; one-component plasma; collective motion in liquids

1. Introduction

About 40 years ago, Robert Zwanzig published an influential paper on the relation between self-diffusion and viscosity of liquids (Stokes–Einstein relation) [1]. The purpose of the present paper is to demonstrate that the dynamical picture behind Zwanzig's result is equivalent to a random walk process, with well defined length and time scales. It is also demonstrated that a theoretical prediction for the numerical factor relating the self-diffusion and viscosity coefficients, in the form of the Stokes–Einstein relation, is quite sensitive to concrete assumptions about the liquid collective mode spectrum. The results provide a consistent picture of the diffusion mechanism in dense liquids with soft isotropic pairwise interactions.

2. Results

2.1. Diffusion as Random Walk

Self-diffusion usually describes the displacement of a test particle immersed in a medium with no external gradients. A canonical example is the Brownian motion, representing a random motion of macroscopic particles suspended in a liquid or a gas. Here, we are interested in atomic scales and, hence, consider displacements of a labeled atom in a fluid of unlabeled, but otherwise identical, atoms. If this motion can be considered a random walk process, then the diffusion coefficient in three spatial dimensions can be defined as [2]

$$D = \frac{1}{6}\frac{\langle r^2 \rangle}{\tau}, \qquad (1)$$

where r is an actual (variable) length of the random walk, τ is the time scale, and we focus on sufficiently long times ($t \gg \tau$). Consider first an ideal gas as an appropriate example. The atoms move freely between pairwise collisions. If the distribution of free paths between collisions follows the $e^{-r/\lambda}/\lambda$ scaling, then $\langle r \rangle = \lambda$ and $\langle r^2 \rangle = 2\lambda^2$, where λ is the mean free path [2]. Combining this with the relation for the average atom velocity $\langle v \rangle = \lambda/\tau$, we recover the elementary kinetic formula for the diffusion coefficient of an ideal gas

$$D = \frac{1}{3}\langle v \rangle \lambda. \qquad (2)$$

The dynamical picture is very different in liquids and this simple consideration clearly does not apply. The very concept of random walk, however, remains relevant, although

characteristic length and time scales associated with a random walk process in liquids are very different from those in gases.

Below, the model proposed by Zwanzig [1] to describe relations between the self-diffusion and shear viscosity coefficients of liquids, is discussed in some detail. In doing so, we naturally repeat some arguments and formulas from Zwanzig's original work and later publications (for instance, from a recent paper by the present author [3]). The emphasis is, however, not on the Stokes–Einstein relation per se, but rather on the possibility of presenting self-diffusion as a random walk process, and on defining the associated length and time scales. The emerging picture represents an alternative to the often assumed hopping mechanism of diffusion in the liquid state.

Zwanzig's approach is based on the assumption that atoms in liquids exhibit solid-like oscillations about temporary equilibrium positions corresponding to a local minimum on the system's potential energy surface [2,4]. These positions do not form a regular lattice like in crystalline solids. They are also not fixed, and change (or drift) with time (this is why liquids can flow), but on much longer time scales. Local configurations of atoms are preserved for some time until a fluctuation in the kinetic energy allows rearranging the positions of some of the atoms towards a new local minimum in the multidimensional potential energy surface. The waiting time distribution of the rearrangements scales as $\exp(-t/\tau)/\tau$, where τ is a lifetime. Atomic motions after the rearrangements are uncorrelated with motions before rearrangements [1].

Within this ansatz, a simplest reasonable approximation for the velocity autocorrelation function of an atom j is

$$Z_j(t) \simeq \left(\frac{T}{m}\right) \cos(\omega_j t) e^{-t/\tau}, \qquad (3)$$

corresponding to a time dependence of a damped harmonic oscillator. Here, T is the temperature in energy units, m is the atomic mass, and ω_j is an effective vibrational frequency. The self-diffusion coefficient D is given by the Green–Kubo formula

$$D = \frac{1}{N} \int_0^\infty \sum_j Z_j(t) dt. \qquad (4)$$

Zwanzig then assumed that vibrational frequencies ω_j are related to the collective mode spectrum and performs averaging over collective modes. After the evaluation of the time integral, this yields

$$D = \frac{T}{3mN} \sum_k \frac{\tau}{1 + \omega_k^2 \tau^2}, \qquad (5)$$

where the summation runs over $3N$ normal mode frequencies. The dynamical picture used by Zwanzig makes sense only if the waiting time τ is much longer than the inverse characteristic frequency of the solid-like oscillations. In this case, we can rewrite Equation (5) as

$$D = \frac{T}{m\tau} \left\langle \frac{1}{\omega^2} \right\rangle, \qquad (6)$$

where the conventional definition of averaging, $\langle \omega^{-2} \rangle = (1/3N) \sum_k \omega_k^{-2}$ has been used.

Equation (6) allows for a simple physical interpretation. It represents a diffusion coefficient for a random walk process, Equation (1). The length scale of this process is identified as

$$\langle r^2 \rangle = \frac{6T}{m} \left\langle \frac{1}{\omega^2} \right\rangle, \qquad (7)$$

which is twice the mean-square displacement of an atom from its local equilibrium position due to solid-like vibrations [5]. The coefficient of two appears, because the initial atom position is not at the local equilibrium, but randomly distributed with the same properties as the final one (after the waiting time τ). The characteristic time scale of the random walk process is just the waiting time τ. Moreover, this waiting time should be associated with the Maxwellian shear relaxation time [2,6]

$$\tau_M = \frac{\eta}{G_\infty} = \frac{\eta}{mnc_t^2}, \qquad (8)$$

where η is the shear viscosity coefficient, G_∞ is the infinite frequency (instantaneous) shear modulus, n is the density, and c_t is the transverse sound velocity.

Thus, self-diffusion in the liquid state can be viewed as a random walk due to atomic vibrations around temporary equilibrium positions over time scales associated with re-arrangements of these equilibrium positions. In this paradigm, consecutive changes of temporary equilibrium positions (jumps of liquid configurations between two neighboring local minima of the multidimensional potential energy surface in Zwanzig's terminology) are relatively small, much smaller than the vibrational amplitude. Hopping events with displacement amplitudes of the order of interatomic separation may be present, but they are relatively rare and do not contribute to the diffusion process. This picture is very different from the widely accepted hopping mechanism of self-diffusion in liquids. Previously, the concept of random walk was suggested in the context of molecular and atomic motion in water and liquid argon [7]. Here, we provide a more quantitative basis for this treatment.

Substituting Equation (8) into Equation (6), we obtain a relation between the self-diffusion and viscosity coefficients in the form of the Stokes–Einstein (SE) relation,

$$D\eta \left(\frac{\Delta}{T} \right) = \frac{c_t^2}{\Delta^2} \left\langle \frac{1}{\omega^2} \right\rangle = \alpha_{SE}, \qquad (9)$$

where $\Delta = n^{-1/3}$ is the mean interatomic separation and α_{SE} is the SE coefficient.

Formula (9) particularly emphasizes the relation between the liquid transport and collective mode properties. Since the exact distribution of frequencies is generally not available, Zwanzig originally used a Debye approximation, characterized by one longitudinal and two transverse modes with acoustic dispersion. The sum over frequencies can be converted to an integral over k using the standard procedure $\sum_k \to V \int d\mathbf{k}/(2\pi)^3$, where V is the volume. This yields

$$\left\langle \frac{1}{\omega^2} \right\rangle = \frac{1}{6\pi^2 n} \int_0^{k_{max}} k^2 dk \left(\frac{1}{\omega_l^2} + \frac{2}{\omega_t^2} \right), \qquad (10)$$

where the cutoff $k_{max} = (6\pi^2 n)^{1/3}$ is chosen to provide n modes in each branch of the spectrum. This ensures that the averaging procedure applied to a quantity that does not depend on k does not change its value. Substituting $\omega_l = c_l k$ and $\omega_t = c_t k$ into Equation (10) we arrive at

$$\alpha_{SE} = \frac{2}{(6\pi^2)^{2/3}} \left(1 + \frac{c_t^2}{2c_l^2} \right) \simeq 0.13 \left(1 + \frac{c_t^2}{2c_l^2} \right). \qquad (11)$$

This essentially coincides with Zwanzig's original result, except he expressed the SE coefficient in terms of the longitudinal and shear viscosity $\alpha_{SE} \simeq 0.13(1 + \eta/2\eta_l)$. The equivalence was pointed out in Reference [6]. Note that since the sound velocity ratio c_t/c_l is confined in the range from 0 to $\sqrt{3}/2$, the coefficient α_{SE} can vary only between $\simeq 0.13$ and $\simeq 0.18$ [1,6]. Possible relations between the viscosity and thermal conductivity coefficients of dense fluids that can complement the SE relations of Equations (9) and (11) have been discussed recently [8].

An important time scale of a liquid state is a structure relaxation time. This can be defined as an average time it takes an atom to move the average interatomic distance Δ

(sometimes it is referred to as the Frenkel relaxation time [9–11]). Taking into account diffusive atomic motions, we can write $\tau_R = \Delta^2/6D$. From Equation (1), we immediately get

$$\tau_R = \frac{\Delta^2}{\langle r^2 \rangle} \tau_M. \tag{12}$$

This implies that $\tau_R/\tau_M \gg 1$. The time scale ratio τ_R/τ_M has a maximum at melting conditions, where, according to the Lindemann melting criterion $\Delta^2/\langle r^2 \rangle \sim 100$ [5,12]. This picture is consistent with the results from numerical simulations (see, e.g., Figure 3 from Reference [11]). Thus, there is a huge separation between the structure relaxation and individual atom dynamical relaxation time scales.

2.2. Relation to Collective Modes Properties

Despite the simplifications involved, the predictive power of Zwanzig's model is quite impressive. Although the model does not allow making independent theoretical predictions of viscosity and self-diffusion coefficients, its prediction of the product, in the form of Equation (9), is highly accurate in some vicinity of the liquid–solid phase transition of many simple liquids [6,13,14]. Moreover, the coefficient α_{SE} can be correlated with the potential softness (via the ratio of the sound velocities), as the model predicts [6]. Some of the assumptions, such as the effect of the waiting time distribution, were critically examined in Reference [15]. In particular, it was demonstrated that the SE relation of the form (9) is not obeyed if the distribution of waiting times is not exponential. In this section, we address another interesting question: how sensitive is the value of α_{SE} to the assumptions about liquid collective mode properties?

To be specific, we consider a model one-component plasma (OCP) system. The OCP fluid is chosen for the following three main reasons: (i) vibrational (caging) motion is most pronounced due to extremely soft and long-ranged character of the interaction potential [16,17]; (ii) Zwanzig's original derivation is not directly applicable to the OCP case, because the longitudinal mode is not acoustic (but plasmon) and, thus, it is a good opportunity to examine how the model should be modified in this case; (iii) collective modes in the OCP system are well studied and understood (for example, simple analytical expressions for the long-wavelength dispersion relations are available, see Appendix A).

The OCP model is an idealized system of mobile point charges immersed in a neutralizing fixed background of opposite charge (e.g., ions in the immobile background of electrons or vice versa) [18–24]. From the fundamental point of view, OCP is characterized by a very soft and long-ranged Coulomb interaction potential, $\phi(r) = q^2/r$, where q is the electric charge. The particle–particle correlations and thermodynamic properties of the OCP are characterized by a single dimensionless coupling parameter $\Gamma = q^2/aT$, where $a = (4\pi n/3)^{-1/3}$ is the Wigner–Seitz radius. At $\Gamma \gtrsim 1$, the OCP is strongly coupled, and this is where it exhibits properties characteristic of a fluid phase (a body centered cubic phase becomes thermodynamically stable at $\Gamma \gtrsim 174$, as the comparison of fluid and solid Helmholtz free energies predicts [22,25,26]). Dynamical scales of the OCP are usually expressed by the plasma frequency $\omega_p = \sqrt{4\pi q^2 n/m}$. For example, the Einstein frequency is $\Omega_E^2 \equiv \langle \omega^2 \rangle = \omega_p^2/3$. The transverse sound velocity at strong coupling is $c_t^2 = (3/100\pi)(4\pi/3)^{1/3}\omega_p^2\Delta^2 \simeq 0.015\omega_p^2\Delta^2$ [27].

From extensive molecular dynamics simulations, it is known that the SE relation is satisfied to a very high accuracy in a strongly coupled OCP fluid with $\alpha_{SE} \simeq 0.14 \pm 0.01$ at $\Gamma \gtrsim 50$ [14,28,29]. Figure 1 demonstrates that α_{SE} approaches the strongly coupled asymptote already at $\Gamma \simeq 10$. Note that the OCP value of the SE coefficient is not truly universal, but rather representative for soft long-ranged pairwise interactions, in which case the transverse-to-longitudinal sound velocity ratio is small [see Equation (11)]. For example, the same value ($\simeq 0.14$) is reached in weakly screened Coulomb (Yukawa) fluids, while for Lennard-Jones fluids it increases to $\alpha_{SE} \simeq 0.15$ and further to $\alpha_{SE} \simeq 0.17$ in hard-sphere fluids [14].

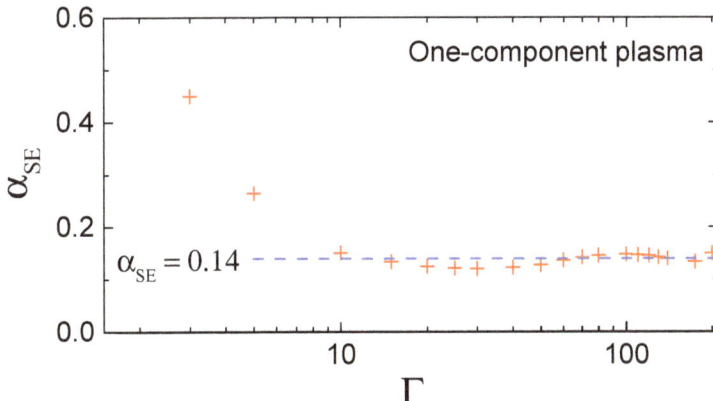

Figure 1. (Color online) Stokes–Einstein parameter α_{SE} as a function of the coupling parameter Γ for a OCP fluid. The symbols correspond to MD simulation results from Refs. [28,29]. The dashed line shows a strong coupling asymptote $\alpha_{SE} \simeq 0.14$.

Now, we examine the sensitivity of the theoretical value of the SE coefficient α_{SE} to concrete assumptions about the collective excitation spectrum. We start with the simplest approximation that all atoms are oscillating with the same Einstein frequency Ω_E (known as the Einstein model in the solid state physics). This approximation results in $\alpha_{SE} \simeq 0.046$, which is too low compared to the actual value from MD simulations (see Figure 1).

As a next level of approximation a Debye-like vibrational density of states (VDOS), $g(\omega) \propto \omega^2$ is assumed (averaging is performed using a standard definition $\langle \omega^\ell \rangle = \int \omega^\ell g(\omega) d\omega / (\int g(\omega) d\omega)$). Using the cutoff Debye frequency ω_D and requesting that $\langle \omega^2 \rangle = \Omega_E^2$ we arrive at $\langle \omega^{-2} \rangle = 9/5\Omega_E^2$. This yields $\alpha_{SE} \simeq 0.083$, which is somewhat better, but still considerably smaller than the actual result.

The most accurate theoretical estimate would be obtained if the exact VDOS were known. However, this is not the case. Nevertheless, accurate knowledge of the real dispersion relations for the longitudinal and transverse modes can be already quite useful. We make use of simple expressions based on the quasi-localized charge approximation (QLCA) [30] combined with the excluded cavity model for the radial distribution function [27]. The corresponding expressions for $\omega_l(k)$ and $\omega_t(k)$ are provided in the Appendix A. Substituting these in Equation (10), we have obtained $\langle \omega_p^2/\omega^2 \rangle \simeq 9.76$ and $\alpha_{SE} \simeq 0.150$. This is very close to the exact result from MD simulations, as expected. Note that the exact result $\langle \omega^2/\omega_p^2 \rangle = 1/3$ is reproduced by construction.

The last demonstration uses a heuristic VDOS of the form

$$g(\omega) = \mathcal{A}\omega^2 \exp(-\mathcal{B}\omega^2), \tag{13}$$

which reproduces the Debye model at low frequencies and implements the Gaussian cutoff at high ω. This form was inspired by the observation that the functional form $g(\omega) = 2\alpha\omega e^{-\alpha\omega^2}$ can fit the numerically obtained VDOS of Lennard-Jones liquids reasonably well [31]. We just substituted the linear scaling at low frequencies with the quadratic one to make the integral converging. This is clearly not a valid physical argument, but we use it here merely for illustrative purposes. The two normalization conditions yield

$$\mathcal{A} = \frac{4}{\sqrt{\pi}} \left(\frac{3}{2\Omega_E^2} \right)^{3/2}, \qquad \mathcal{B} = \frac{3}{2\Omega_E^2}. \tag{14}$$

Application of this VDOS results in $\alpha_{SE} \simeq 0.139$, which almost coincides with the exact MD result. Thus, implementation of the Gaussian cutoff to the Debye-like VDOS improves the situation considerably.

It should be noted that very long wavelengths and low frequency parts of the spectra are not relevant for the present consideration, because dynamics on time scales shorter than the relaxation time τ_M is considered. However, since $\omega\tau_M \gg 1$ needs to be satisfied, this corresponds to only a small part of the entire spectrum, and we therefore included low frequencies for simplicity, similar to what Zwanzig did originally [1]. This also allows us to disregard the effects associated with the k-gap in the dispersion relation of the transverse mode, an important property of liquid dynamics [32–37].

3. Discussion and Conclusions

While transport phenomena in gaseous and solid phases can be well described at the quantitative level, transport in liquids is still much less understood, even at the qualitative level. Here, we have demonstrated that self-diffusion in dense liquids can be described as a random walk process with well defined time and length scales. The length scale is related to the amplitude of solid-like vibrations around local temporary equilibrium positions. The time scale is set by the Maxwellian shear relaxation time. This dynamical picture results in the Stokes–Einstein relation between the coefficients of self-diffusion and viscosity, which is satisfied in many simple liquids. Importantly, the hoping mechanism of atomic diffusion in liquids is irrelevant in this picture of microscopic atomic dynamics.

The dynamical picture involved requires that the atomic motion be dominated by fast solid-like oscillations around the local equilibrium positions. This limits the model applicability to regions on the phase diagram located not too far from the liquid–solid phase transition (high densities and low temperatures). Additionally, it applies to sufficiently soft interaction potentials with pronounced oscillation dynamics. In the hard sphere interaction limit, this model is clearly inadequate (although SE relation is still satisfied even in this limit [14]).

Finally, we have demonstrated that a theoretically obtained numerical factor in the SE relation is sensitive to concrete assumptions about the liquid collective modes properties. This highlights the necessity of accurate knowledge of the vibrational density of states and dispersion relations in liquids.

Funding: This research was supported by The Ministry of Science and Higher Education of the Russian Federation (agreement with the Joint Institute for High Temperatures RAS No. 075-15-2020-785, dated 23 September 2020).

Institutional Review Board Statement: Not applicable.

Informed Consent Statement: Not applicable.

Data Availability Statement: Not applicable as no new date were created or analyzed in this study.

Conflicts of Interest: The author declares no conflict of interest.

Abbreviations

The following abbreviations are used in this manuscript:

SE relation	Stokes–Einstein relation
OCP	one-component plasma
VDOS	vibrational density of states
QLCA	quasi-localized charge approximation

Appendix A. Dispersion Relations of a Strongly Coupled OCP Fluid

Table A1. Averaged frequencies of a strongly coupled OCP fluid obtained with the help of dispersion relations (A1) and (A2).

$\langle \omega^2/\omega_p^2 \rangle$	$\langle \omega/\omega_p \rangle$	$\langle \ln \omega/\omega_p \rangle$	$\langle \omega_p/\omega \rangle$	$\langle \omega_p^2/\omega^2 \rangle$
$\frac{1}{3}$	0.514	-0.8023	2.5856	9.7623

Combining the QLCA model with a simple excluded cavity approximation for the radial distribution function, the following analytical expressions for the longitudinal and transverse dispersion relations in OCP fluids have been derived [27]

$$\omega_l^2 = \omega_p^2 \left(\frac{1}{3} - \frac{2\cos Rq}{R^2 q^2} + \frac{2\sin Rq}{R^3 q^3} \right) \tag{A1}$$

and

$$\omega_t^2 = \omega_p^2 \left(\frac{1}{3} + \frac{\cos Rq}{R^2 q^2} - \frac{\sin Rq}{R^3 q^3} \right), \tag{A2}$$

where $q = ka$ is the reduced wave-number and R is the reduced excluded cavity radius. In the strongly coupled OCP regime, we have $R = \sqrt{6/5} \simeq 1.09545$. Expressions (A1) and (A2) are rather accurate in the long-wavelength regime [38–40], except the existence of k-gap in the transverse mode is not accounted for [36]. Expressions (A1) and (A2) can be used to perform averaging over collective mode frequencies. We performed averaging of several frequency-related quantities and provide them in Table A1 for completeness.

The result for $\langle \omega^2/\omega_p^2 \rangle \equiv 1/3$ is exact by virtue of Equations (A1) and (A2). The quantity $\langle \omega_p^2/\omega^2 \rangle$ is used here to estimate the SE coefficient. The quantity $\langle \omega/\omega_p \rangle$ emerges in the vibrational model of thermal conductivity of simple fluids [3,41]. The quantity $\langle \ln \omega/\omega_p \rangle$ emerges in a variant of the cell theory of liquid entropy [42].

References

1. Zwanzig, R. On the relation between self-diffusion and viscosity of liquids. *J. Chem. Phys.* **1983**, *79*, 4507–4508. [CrossRef]
2. Frenkel, Y. *Kinetic Theory of Liquids*; Dover: New York, NY, USA, 1955.
3. Khrapak, S.A. Vibrational model of thermal conduction for fluids with soft interactions. *Phys. Rev. E* **2021**, *103*, 013207. [CrossRef]
4. Stillinger, F.H.; Weber, T.A. Hidden structure in liquids. *Phys. Rev. A* **1982**, *25*, 978–989. [CrossRef]
5. Khrapak, S.A. Lindemann melting criterion in two dimensions. *Phys. Rev. Res.* **2020**, *2*, 012040. [CrossRef]
6. Khrapak, S. Stokes–Einstein relation in simple fluids revisited. *Mol. Phys.* **2019**, *118*, e1643045. [CrossRef]
7. Berezhkovskii, A.M.; Sutmann, G. Time and length scales for diffusion in liquids. *Phys. Rev. E* **2002**, *65*, 060201. [CrossRef] [PubMed]
8. Khrapak, S.A.; Khrapak, A.G. Correlations between the Shear Viscosity and Thermal Conductivity Coefficients of Dense Simple Liquids. *JETP Lett.* **2021**, *114*, 540. [CrossRef]
9. Brazhkin, V.V.; Fomin, Y.D.; Lyapin, A.G.; Ryzhov, V.N.; Trachenko, K. Two liquid states of matter: A dynamic line on a phase diagram. *Phys. Rev. E* **2012**, *85*, 031203. [CrossRef] [PubMed]
10. Brazhkin, V.V.; Lyapin, A.; Ryzhov, V.N.; Trachenko, K.; Fomin, Y.D.; Tsiok, E.N. Where is the supercritical fluid on the phase diagram? *Phys.-Uspekhi* **2012**, *182*, 1137–1156. [CrossRef]
11. Bryk, T.; Gorelli, F.A.; Mryglod, I.; Ruocco, G.; Santoro, M.; Scopigno, T. Reply to "Comment on 'Behavior of Supercritical Fluids across the Frenkel Line'". *J. Phys. Chem. B* **2018**, *122*, 6120–6123. [CrossRef]
12. Lindemann, F. The calculation of molecular vibration frequencies. *Z. Phys.* **1910**, *11*, 609.
13. Costigliola, L.; Heyes, D.M.; Schrøder, T.B.; Dyre, J.C. Revisiting the Stokes-Einstein relation without a hydrodynamic diameter. *J. Chem. Phys.* **2019**, *150*, 021101. [CrossRef] [PubMed]
14. Khrapak, S.; Khrapak, A. Excess entropy and the Stokes–Einstein relation in simple fluids. *Phys. Rev. E* **2021**, *104*, 044110. doi:10.1103/PhysRevE.104.044110 [CrossRef] [PubMed]
15. Mohanty, U. Inferences from a microscopic model for the Stokes-Einstein relation. *Phys. Rev. A* **1985**, *32*, 3055–3058. [CrossRef]
16. Donko, Z.; Kalman, G.J.; Golden, K.I. Caging of Particles in One-Component Plasmas. *Phys. Rev. Lett.* **2002**, *88*, 225001. [CrossRef] [PubMed]
17. Daligault, J. Universal Character of Atomic Motions at the Liquid-Solid Transition. *arXiv* **2020**, arXiv:2009.14718.
18. Brush, S.G.; Sahlin, H.L.; Teller, E. Monte Carlo Study of a One-Component Plasma. *J. Chem. Phys.* **1966**, *45*, 2102–2118. [CrossRef]

19. Hansen, J.P. Statistical Mechanics of Dense Ionized Matter. I. Equilibrium Properties of the Classical One-Component Plasma. *Phys. Rev. A* **1973**, *8*, 3096–3109. [CrossRef]
20. DeWitt, H.E. Statistical mechnics of dense plasmas: Numerical simulation and theory. *J. Phys. Colloq.* **1978**, *39*, C1-173–C1-180. [CrossRef]
21. Baus, M.; Hansen, J.P. Statistical mechanics of simple Coulomb systems. *Phys. Rep.* **1980**, *59*, 1–94. [CrossRef]
22. Ichimaru, S. Strongly coupled plasmas: High-density classical plasmas and degenerate electron liquids. *Rev. Mod. Phys.* **1982**, *54*, 1017–1059. [CrossRef]
23. Stringfellow, G.S.; DeWitt, H.E.; Slattery, W.L. Equation of state of the one-component plasma derived from precision Monte Carlo calculations. *Phys. Rev. A* **1990**, *41*, 1105–1111. [CrossRef] [PubMed]
24. Fortov, V.; Iakubov, I.; Khrapak, A. *Physics of Strongly Coupled Plasma*; OUP Oxford: New York, NY, USA; London, UK, 2006.
25. Dubin, D.H.E.; O'Neil, T.M. Trapped nonneutral plasmas, liquids, and crystals (the thermal equilibrium states). *Rev. Mod. Phys.* **1999**, *71*, 87–172. [CrossRef]
26. Khrapak, S.A.; Khrapak, A.G. Internal Energy of the Classical Two- and Three-Dimensional One-Component-Plasma. *Contrib. Plasma Phys.* **2016**, *56*, 270–280. [CrossRef]
27. Khrapak, S.A.; Klumov, B.; Couedel, L.; Thomas, H.M. On the long-waves dispersion in Yukawa systems. *Phys. Plasmas* **2016**, *23*, 023702. [CrossRef]
28. Daligault, J. Practical model for the self-diffusion coefficient in Yukawa one-component plasmas. *Phys. Rev. E* **2012**, *86*, 047401. [CrossRef]
29. Daligault, J.; Rasmussen, K.; Baalrud, S.D. Determination of the shear viscosity of the one-component plasma. *Phys. Rev. E* **2014**, *90*, 033105. [CrossRef]
30. Golden, K.I.; Kalman, G.J. Quasilocalized charge approximation in strongly coupled plasma physics. *Phys. Plasmas* **2000**, *7*, 14–32. [CrossRef]
31. Rabani, E.; Gezelter, J.D.; Berne, B.J. Calculating the hopping rate for self-diffusion on rough potential energy surfaces: Cage correlations. *J. Chem. Phys.* **1997**, *107*, 6867–6876. [CrossRef]
32. Hansen, J.P.; McDonald, I.R. *Theory of Simple Liquids*; Elsevier: Amsterdam, The Netherlands, 2006.
33. Trachenko, K.; Brazhkin, V.V. Collective modes and thermodynamics of the liquid state. *Rep. Progr. Phys.* **2015**, *79*, 016502. [CrossRef]
34. Bolmatov, D.; Zhernenkov, M.; Zav'yalov, D.; Stoupin, S.; Cunsolo, A.; Cai, Y.Q. Thermally triggered phononic gaps in liquids at THz scale. *Sci. Rep.* **2016**, *6*, 19469. [CrossRef] [PubMed]
35. Kryuchkov, N.P.; Mistryukova, L.A.; Brazhkin, V.V.; Yurchenko, S.O. Excitation spectra in fluids: How to analyze them properly. *Sci. Rep.* **2019**, *9*, 10483. [CrossRef]
36. Khrapak, S.A.; Khrapak, A.G.; Kryuchkov, N.P.; Yurchenko, S.O. Onset of transverse (shear) waves in strongly-coupled Yukawa fluids. *J. Chem. Phys.* **2019**, *150*, 104503. [CrossRef]
37. Kryuchkov, N.P.; Mistryukova, L.A.; Sapelkin, A.V.; Brazhkin, V.V.; Yurchenko, S.O. Universal Effect of Excitation Dispersion on the Heat Capacity and Gapped States in Fluids. *Phys. Rev. Lett.* **2020**, *125*, 125501. [CrossRef]
38. Khrapak, S.A. Practical dispersion relations for strongly coupled plasma fluids. *AIP Adv.* **2017**, *7*, 125026. [CrossRef]
39. Khrapak, S.; Khrapak, A. Simple Dispersion Relations for Coulomb and Yukawa Fluids. *IEEE Trans. Plasma Sci.* **2018**, *46*, 737–742. [CrossRef]
40. Fairushin, I.; Khrapak, S.; Mokshin, A. Direct evaluation of the physical characteristics of Yukawa fluids based on a simple approximation for the radial distribution function. *Results Phys.* **2020**, *19*, 103359. [CrossRef]
41. Khrapak, S.A. Thermal conductivity of strongly coupled Yukawa fluids. *Phys. Plasmas* **2021**, *28*, 084501. [CrossRef]
42. Khrapak, S.A.; Yurchenko, S.O. Entropy of simple fluids with repulsive interactions near freezing. *J. Chem. Phys.* **2021**, *155*, 134501. [CrossRef]

Article

MHD Stagnation Point on Nanofluid Flow and Heat Transfer of Carbon Nanotube over a Shrinking Surface with Heat Sink Effect

Mohamad Nizam Othman [1], Alias Jedi [1,2,*] and Nor Ashikin Abu Bakar [3]

[1] Department of Mechanical and Manufacturing Engineering, Faculty of Engineering and Built Environment, Universiti Kebangsaan Malaysia, Bangi 43600, Malaysia; P102090@siswa.ukm.edu.my

[2] Centre for Automotive Research (CAR), Faculty of Engineering and Built Environment, Universiti Kebangsaan Malaysia, Bangi 43600, Malaysia

[3] Institute of Engineering Mathematics, Faculty of Applied and Human Sciences, Universiti Malaysia Perlis, Arau 02600, Malaysia; ashikinbakar@unimap.edu.my

* Correspondence: aliasjedi@ukm.edu.my

Abstract: This study is to investigate the magnetohydrodynamic (MHD) stagnation point flow and heat transfer characteristic nanofluid of carbon nanotube (CNTs) over the shrinking surface with heat sink effects. Similarity equations deduced from momentum and energy equation of partial differential equations are solved numerically. This study looks at the different parameters of the flow and heat transfer using first phase model which is Tiwari-Das. The parameter discussed were volume fraction nanoparticle, magnetic parameter, heat sink/source parameters, and a different type of nanofluid and based fluids. Present results revealed that the rate of nanofluid (SWCNT/kerosene) in terms of flow and heat transfer is better than (MWCNT/kerosene) and (CNT/water) and regular fluid (water). Graphically, the variation results of dual solution exist for shrinking parameter in range $\lambda_c < \lambda \leq -1$ for different values of volume fraction nanoparticle, magnetic, heat sink parameters, and a different type of nanofluid. However, a unique solution exists at $-1 < \lambda < 1$, and no solutions exist at $\lambda < \lambda_c$ which is a critical value. In addition, the local Nusselt number decreases with increasing volume fraction nanoparticle when there exists a heat sink effect. The values of the skin friction coefficient and local Nusselt number increase for both solutions with the increase in magnetic parameter. In this study, the investigation on the flow and heat transfer of MHD stagnation point nanofluid through a shrinking surface with heat sink effect shows how important the application to industrial applications.

Keywords: MHD stagnation flow; nanofluid; heat transfer; carbon nanotube; heat sink

1. Introduction

Currently, nanofluid plays an important role in heat transfer enhancement. This is due to the efficiency of heat transfer, and it is useful in most components such as heat exchanger, electronic devices, and any equipment that involve on heat transfer rate. Conventional heat transfer fluid or base fluid such as water, kerosene, oil, and ethylene glycol have a low heat transfer rate due to poor thermal conductivity. Therefore, this shortcoming of heat transfer performance can be overcome by adding a single type of nanosized particle into base fluid. That is why nanofluid research has still been relevant in engineering and industrial application until today. Initially, study about heat transfer characteristics of nanofluid is reviewed [1]. It already mentions that convective heat transfer can be enhanced passively by enhancing thermal conductivity of the fluid. Next, Asirvatham et al. [2] investigated convective heat transfer of nanofluid with correlations. N. Kumar et al. [3] studied on nanofluid application for heat transfer in a microchannel. Numerical study of convective heat transfer of nanofluid is reviewed by Vanaki et al. [4]. It indicated that effective thermal conductivity and viscosity of nanofluid are predicted by considering the effect of volume fraction, particle shape, particle size, nanofluid temperature, and Brownian diffusion. Han et al. [5] conducted the experimental study of heat transfer enhancement using nanofluid in a double-tube heat exchanger. They concluded that heat transfer at boundary layer increases significantly with the addition of nanoparticles as constant bombarding of nanosized particle transfers much of the heat from the boundary to the mainstream fluid, thus increasing the heat transfer effect and Nusselt number. Furthermore, Chiam et al. [6] presented the numerical study of nanofluid heat transfer for different tube geometries. They mentioned that the convective heat transfer coefficient is strongly dependent on the surface of the solid, thermophysical properties of coolant, and the type of flow. Ahmadi and Willing [7] studied the heat transfer measurement in water based nanofluid. The study about flow and heat transfer behavior of nanofluid in microchannels is investigated by Bowers et al. [8]. They mentioned that the nanoparticles need to be as stable as possible to avoid clogging and sedimentation within heat transfer equipment. The study of Buschmann et al. [9] about the correct interpretation of nanofluid convective heat transfer has proven that the heat transfer enhancement provided by nanofluid equals the increase in the thermal conductivity of the nanofluid as compared to the base fluid that is independent of the nanoparticle concentration or material.

This study also involved the stagnation flow toward the shrinking sheet is already conducted by [10], which described the fluid motion near the stagnation region, which exists on all solid bodies moving in a fluid. This region encounters the highest pressure, heat transfer, and rates of mass deposition. Bhatti et al. [11] conducted the numerical simulation of fluid flow over a shrinking porous sheet by successive linearization method. This study confirmed the existence of a dual solution for shrinking sheet, while for the stretching case, the solution is unique. Soid et al. [12] investigated the axisymmetric stagnation-point of second-order velocity slip. It mentions that the value of skin friction coefficient being zero when $\lambda = 1$, because the fluid and the solid surface which move in the same velocity, and thus, there is no friction at the fluid-solid interface. However, there is a heat transfer at the surface, even though no friction occurred. This happens because of the temperature difference between the fluid and the solid surface. Dash et al. [13] presented the numerical approach to boundary layer stagnation-point flow past a stretching/shrinking sheet. It explained that the striking feature of the observation is that the shrinking of the boundary surface overrides the resistive effect of the electromagnetic force and sustains a backflow. Tasawar Hayat et al. [14] considered the inclined magnetic field and heat source/sink aspects in flow of nanofluid with nonlinear thermal radiation and examined the numerical simulation for melting heat transfer and radiation effects in stagnation point flow of carbon-water nanofluid [14].

The entropy generation on MHD flow and convective heat transfer in a porous medium of exponentially stretching surface saturated by nanofluids [15]. The study indicated that there is no viscous effect at the exterior to the boundary layer, and therefore, the pressure distribution can be obtained through the Euler form of the momentum equation. The thermal boundary layer in stagnation-point flow past a permeable shrinking sheet with variable surface temperature was studied by Uddin and Bhattacharyya [16]. The stagnation point flow of a micropolar nanofluid past a circular cylinder with velocity and thermal slip was explored by Abbas et al. [17]. The hydromagnetic unsteady slip stagnation flow of nanofluid with suspension of mixed bioconvection was investigated by R. Kumar et al. [18]. They discussed that region $\lambda = 0$ depicts that these forces are of equal magnitude. Mustafa et al. [19] considered an analytical solution of least square method. It is observed that the range of the dual solutions become larger by enhancing the effects of magnetic parameter. Al-Amri and Muthtamilselvan [20] examined that the stagnation point flow of nanofluid containing micro-organisms. Anuar et al. [21] investigated the MHD flow past a nonlinear stretching/shrinking sheet in carbon nanotubes including stability analysis. It is clearly that when $\lambda < \lambda_c$, the solution does not exist because boundary layer separation occurs that causes the boundary layer equation to be invalid. The finding was also similar found by [22]. Furthermore, the study of stretching/shrinking sheet of magnetic nanofluid might be helpful for researchers to study the stability of working fluid [23,24].

In a nutshell, this article is considered MHD stagnation point nanofluid flow and heat transfer of carbon nanotube over a shrinking surface with heat sink effect. The molecular interaction of SWCNT considering the stagnation point and heat sink in different based fluids has few studies discussed by many researchers. Due to high thermal conductivity of CNT and the potential to improve heat transfer, thus, this study also considers water and kerosine as a base fluid and carbon nanotube (CNT) including single-wall carbon nanotube (SWCNT) and multiwall carbon nanotube (MWCNT) as a nanoparticle. The results were obtained in numerical, and they are presented in form of graphs and tables to describe the behavior of this study and are compared with previously published results to achieve a good agreement.

2. Methodology

Let continuity Equation (1), momentum Equation (2), and energy Equation (3) be

$$\frac{\partial u}{\partial x} + \frac{\partial v}{\partial y} = 0 \tag{1}$$

$$u\frac{\partial u}{\partial x} + v\frac{\partial u}{\partial y} = \frac{\mu_{nf}}{\rho_{nf}}\left[\frac{\partial^2 u}{\partial y^2}\right] + U_\infty \frac{\partial U_\infty}{\partial x} + \frac{\sigma B_0^2}{\rho_{nf}}(U_\infty - U) \tag{2}$$

$$u\frac{\partial T}{\partial x} + v\frac{\partial T}{\partial y} = \frac{k_{nf}}{(\rho C_p)_{nf}}\frac{\partial^2 T}{\partial y^2} + \frac{Q}{(\rho C_p)_{nf}}(T - T_\infty) \tag{3}$$

given that the boundary condition of governing equation is given as follow:

$$u = U_w = cx,\ v = 0,\ T = T_w \text{ as } y = 0\quad u = U_\infty = ax,\ T = T_\infty \text{ as } y \to \infty \tag{4}$$

where (u, v) is velocity component along x and y axis, respectively. U_w as velocity wall, U_∞ is free flow velocity, T is temperature, and T_∞ is ambient temperature. (a, c) is a positive constant which refer to stretching/shrinking strength where stretching case is $c > 0$ whereas shrinking case is $c < 0$, and T_w is temperature wall. All nomenclature in Equations (1)–(3) are illustrated in Nomenclature. Figure 1 show the working flow and heat transfer for MHD stagnation point with shrinking surface. For this case, the assumptions of impermeable wall, uniform nanoparticles size, agglomeration effect, and viscous dissipation are neglected. The base fluid and the nanoparticles are similarly considered to be in thermal equilibrium in the Tiwari–Das nanofluid model, with no-slip between them. Heat transport, convection, and the heat sink effect are all accounted for in energy equations. This study focuses on laminar flow for the working liquid; hence, it is expected that large velocity gradient existed, and therefore, the viscous dissipation term in Equation (3) is omitted.

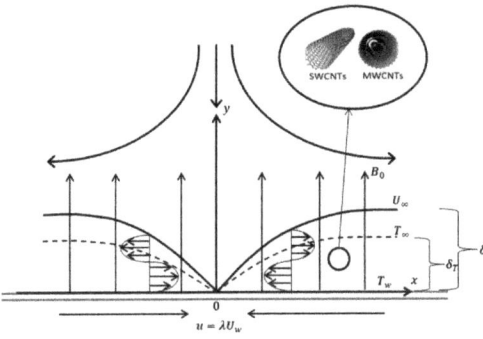

Figure 1. MHD stagnation flow of nanofluid past shrinking sheet.

The nanoparticle used is considered in this problem study to discover the behavior the MHD flow and heat transfer of nanofluid. Therefore, Table 1 shows that the effective thermophysical properties of nanofluid needed to explain the nanofluid model. The empirical shape factor is set $n = 3/m = 3$ where m is referred to ideal spherical shape. Table 2 shows the thermophysical properties used by [15,21] for different nanoparticle and fluid selected in this problem study. In this method, the authors investigate the outcome for flow and heat transfer simultaneously using bvp4c. Hence, the influence of carbon nanotube aspect ratio is neglected.

Table 1. The effective thermophysical properties of nanofluid.

Thermophysical Properties	Nanofluid CNT-Water (s = CNT: n = 3)
Density (kg/m^3)	$\rho_{nf} = (1-\varphi)\rho_f + \varphi\rho_s$
Heat capacity (J/K)	$(\rho C_p)_{nf} = (1-\varphi)(\rho C_p)_f + \varphi(\rho C_p)_s$
Viscosity (Ns/m^{-2})	$\mu_{nf} = \frac{\mu_f}{(1-\varphi)^{2.5}}$
Thermal conductivity (W/Km)	$\frac{k_{nf}}{k_f} = \frac{k_s+(n-1)k_f-(n-1)\varphi(k_f-k_s)}{k_s+(n-1)k_f+\varphi(k_f-k_s)}$

Table 2. Thermophysical properties of nanoparticle, nanoparticle, and base fluid [15,21].

Thermophysical Properties	Base Fluid		Nanoparticle	
	Water Pr = 6.2	Kerosine Pr = 21	SWCNT	MWCNT
ρ (kg/m^3)	997.1	783	2600	1600
C_p ($J/kg\ K$)	4179	2090	425	796
k (W/mK)	0.613	0.145	6600	3000

Furthermore, stagnation point flow in this problem study was also considered. This is because stagnation point flow produced on static surface either stretching or shrinking. Therefore, the similarity solution for the problem of MHD stagnation point flow in nanofluid and heat transfer over shrinking surface with heat sink effect is given as follows:

$$\psi = \sqrt{\nu_f\, ax}\, f(\eta),\quad \eta = \sqrt{\frac{a}{\nu_f}}\, y,\quad \theta(\eta) = \frac{T-T_\infty}{T_w-T_\infty} \tag{5}$$

with ψ being a stream function that is defined as u and v. Thus,

$$u = \frac{\partial \psi}{\partial y} = axf'(\eta),\quad v = -\frac{\partial \psi}{\partial x} = -\sqrt{a\nu_f}\, f(\eta) \tag{6}$$

where (') shows the differentiation with respect to η. Thus, the mathematical model in form of ordinary differential equation (ODE) is stated as follows:

$$\frac{1}{(1-\varphi)^{2.5}} f'''(\eta) - \frac{\rho_{nf}}{\rho_f}\left(f'^2(\eta) - f(\eta)f''(\eta) - 1\right) + M(1-f'(\eta)) = 0$$
$$\frac{1}{Pr}\frac{k_{nf}}{k_f}\theta''(\eta) + \left((1-\varphi) + \varphi\frac{(\rho C_p)_s}{(\rho C_p)_f}\right) f(\eta)\theta'(\eta) + \varepsilon\theta(\eta) = 0 \tag{7}$$

subject to boundary condition:

$$f'(\eta) = \lambda,\ f(\eta) = 0,\ \theta(\eta) = 1 \text{ as } \eta = 0$$
$$f'(\eta) = 1,\ \theta(\eta) = 0 \text{ as } \eta \to \infty \tag{8}$$

where (') represent differentiation with respect to η, φ as volume fraction nanoparticle, ρ_{nf} is density of nanofluid, ρ_f is density of fluid, $M = \sigma B_0^2/a\rho_f$ is magnetic parameter, $Pr = \nu_f/\alpha_f$ as Prandtl number, $(\rho C_p)_{nf}$ is specific heat capacity of nanofluid, $(\rho C_p)_f$ is specific heat capacity of fluid, and $\varepsilon = Q/a(\rho C_p)_f$ is heat sink/source parameter. As for the boundary condition, $\lambda = c/a$ is referred to stretching/shrinking strength or velocity parameter, where $\lambda > 0$ is stretching case, whereas $\lambda < 0$ is shrinking case. The interpretation of the physical quantity considered in the study are local skin friction coefficient, C_f and local number Nusselt, Nu_x which can be given as follows:

$$C_f = \frac{\tau_w}{\rho_f U_\infty^2}, \quad Nu_x = \frac{xq_w}{k_f(T_w - T_\infty)} \tag{9}$$

with shear stress, τ_w and heat flux, q_w which can be defined as follows:

$$\tau_w = \mu_{nf}\left(\frac{\partial u}{\partial y}\right)_{y=0}, \quad q_w = -k_{nf}\left(\frac{\partial T}{\partial y}\right)_{y=0} \tag{10}$$

By using the Reynold number coefficient, $Re_x = U_\infty x/v_f$, thus local skin friction coefficient, $Re_x^{1/2}C_f$ and local number Nusselt, $Re_x^{-1/2}Nu_x$ can be stated as follows:

$$Re_x^{1/2}C_f = \frac{1}{(1-\varphi)^{2.5}}f''(0)$$
$$Re_x^{-1/2}Nu_x = -\frac{k_{nf}}{k_f}\theta'(0) \tag{11}$$

In this study, we investigate the MHD laminar flow where the condition of the flow is assumed stable. Hence, we do not consider the stability analysis for the first and second solutions. Throughout Equations (1)–(11), the Tiwari–Das nanofluid model did not consider the mass transfer of carbon nanotubes. However, the formulation of nanoparticle volume fraction is the advantage of this model to explain the interaction of nanoparticle with working fluids. For this case, the flow and heat transfer of nanofluid are produced and present numerically. Although recent studies [25,26] investigated the nonuniform dispersion of nanoparticle, the Tiwari–Das would be able to measure the flow and heat transfer of nanofluids. However, the limitation of this method is not being able to measure the Brownian motion and thermophoresis interaction between nanoparticles. As described by [27], Brownian dynamic might be able to measure the nonuniform dispersion of nanoparticle.

3. Results

The numerical solutions from the governing ordinary differential equation for flow and energy with its boundary condition were solved by using bvp4c solver in MATLAB software. This solver is based on three-stage collocation at Lobatto point which means the three-stage Lobatto IIIA method. Lobatto IIIA methods can be very efficient for the numerical solution of nonlinear stiff systems (12). These numerical solutions are analyzed and presented in tables and graphs for discussing the behavior of flow and heat transfer of this boundary layer model when including a few parameters. This study is conducted by adding nanoparticle volume fraction of CNT from 0 to 0.2 in range $0 \leq \varphi \leq 0.2$ into base fluid which are water and kerosene selected. Besides that, the parameters values which varied on λ are φ, M, ε, and the nanofluid selected as well as this parameter value varied in region $\lambda_c < \lambda < 1$. This is because second solution is discovered in range $\lambda < -1$ that shows shrinking case and meet the requirements of the study conducted. Next, few numerical results produced of $f''(0)$ are compared with previous results that are shown in Table 3.

Table 3. Comparison several of numerical result $f''(0)$ when stretching/shrinking case and $M = \varepsilon = 0$ with the change of φ and λ in nanofluid.

φ	λ	$f''(0)$					
		Present Result		Bachok et al. [28]		Wang [29]	
		CNT-Water		Cu-Water		Water	
		First Solution	Second Solution	First Solution	Second Solution	First Solution	Second Solution
0	2	−1.887306668		−1.887307		−1.88731	
	1	0		0		0	
	0.5	0.71329495		0.713295		0.7133	
	0	1.232587647		1.232588		1.232588	
	−0.5	1.495669739		1.49567		1.49567	
	−1	1.328816861	0	1.328817	0	1.32882	0
	−1.2	0.932473188	0.233649469	0.932473	0.23365		
0.1	2	−1.78244491		−2.217106			
	1	0		0			
	0.5	0.673663145		0.83794			
	0	1.164103115		1.447977			
	−0.5	1.412567954		1.757032			
	−1	1.254985673	0	1.561022	0		
	−1.2	0.880663529	0.220667553	1.095419	0.274479		
0.2	2	−1.641498824		−2.298822			
	1	0		0			
	0.5	0.620393516		0.868824			
	0	1.072052147		1.501346			
	−0.5	1.300869722		1.821791			
	−1	1.155748188	0	1.618557	0		
	−1.2	0.811025466	0.20321847	1.135794	0.284596		

Based on Table 3, it was found that each numerical result is compared to achieve a good agreement when the nanoparticle is not considered in base fluid. The numerical result obtained is compared with previous result to ensure mathematical model developed and solver method used are valid before numerical solution when the set up for the parameter selected is produced. However, the present comparison results numerically for the $f''(0)$ value with [28] the results being slightly different when volume fraction nanoparticle, and φ is added in base fluids, which are 0.1 and 0.2. In this study, the range of nanoparticle volume fraction with a range of 0–0.2 is chosen based on study by [29], whereas the range of magnetic parameters is set between 0 and 0.2. The numerical result is different because the nanoparticle used is different in base fluid. The problem in this study with the nanoparticle selected is the carbon nanotube (CNT), whereas for Bachok's (2011) study, it was copper. Therefore, different types of nanoparticles in same base fluid have different thermophysical properties and of course give the different numerical result of $f''(0)$ and $-\theta'(0)$.

Figures 2–6 shows the existence of a dual solution clearly. This solution can be observed in region $\lambda_c \leq \lambda \leq 1$ and the existence of unique solution at point $\lambda_c = \lambda$ where λ_c is critical point as well as at region $\lambda > -1$. Based on region produced $\lambda_c \leq \lambda \leq 1$, the mathematical model developed has the potential to describe the behavior of MHD nanofluid over the different parameter values. A numerical solution does not exist when in the region $\lambda_c > \lambda$. Thus, this case shows the incompatibility mathematical model in the region or not being able to easily understand the boundary layer separation and boundary layer approximation are physically cannot be realized. The discussion of this problem study is continued with addition of carbon nanotube (CNT), which is a single-wall carbon nanotube (SWCNT) on local skin friction coefficient, $Re_x^{1/2}C_f$ and local Nusselt number, $Re_x^{-1/2}Nu_x$. Figure 2a indicates the change in trend $Re_x^{1/2}C_f$ which can be referred to as $f''(0)$ on the variation value of volume fraction nanoparticle of SWCNT, φ when stretching/shrinking surface. Stretching/shrinking case that shows in first solution describe the reduction of $Re_x^{1/2}C_f$ when the value of φ increases from 0 to 0.2. Although nanofluid becomes more viscous, it is still not enough to achieve the enhancement of $Re_x^{1/2}C_f$ when the value of φ increases. Therefore, the enlargement in momentum boundary layer thickness, δ which coincides with the increase in value of φ that causes the decrease in $Re_x^{1/2}C_f$ when the value of φ increases.

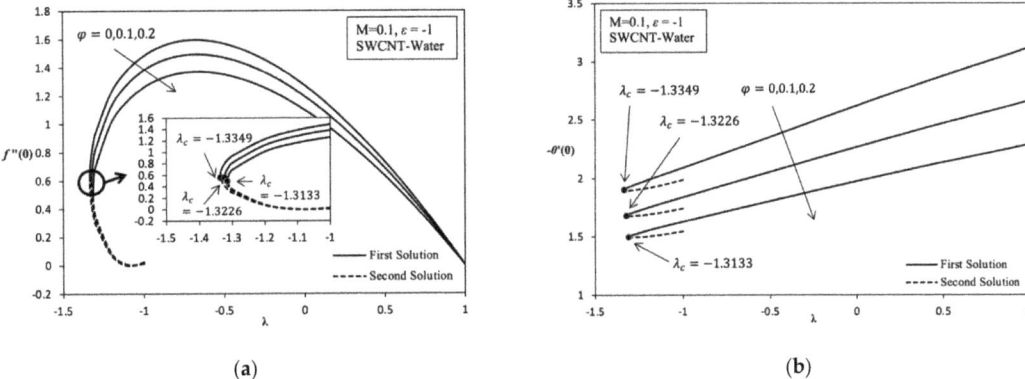

Figure 2. Variation on λ with various values of φ in SWCNT-water when $M = 0.1$ and $\varepsilon = -1$: (**a**) Variation on $f''(0)$; (**b**) Variation on $-\theta\prime(0)$.

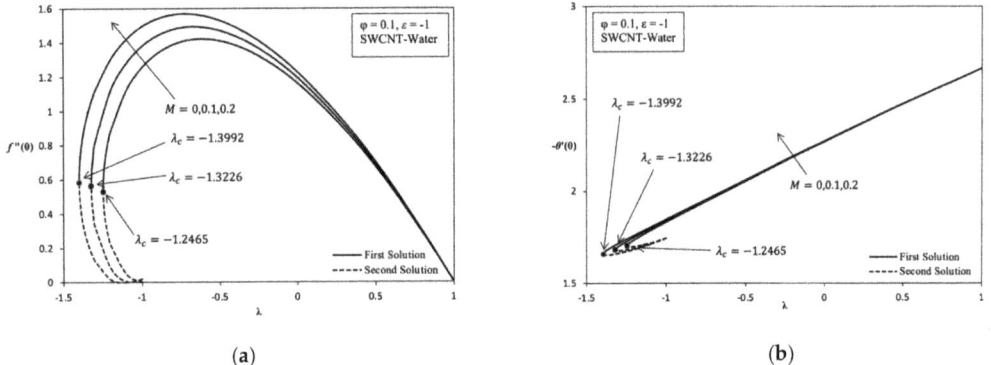

Figure 3. Variation on λ with various values of M in SWCNT-water when $\varphi = 0.1$ and $\varepsilon = -1$: (**a**) Variation on $f''(0)$; (**b**) Variation on $-\theta\prime(0)$.

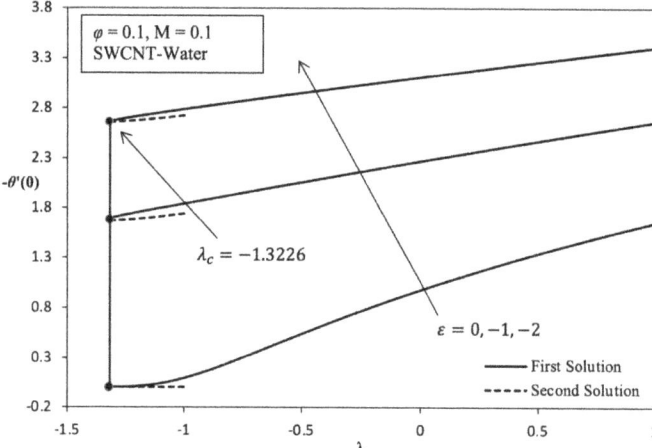

Figure 4. Variation of $-\theta'(0)$ versus λ with various values of ε in SWCNT-water when $\varphi = 0.1$ and $M = 0.1$.

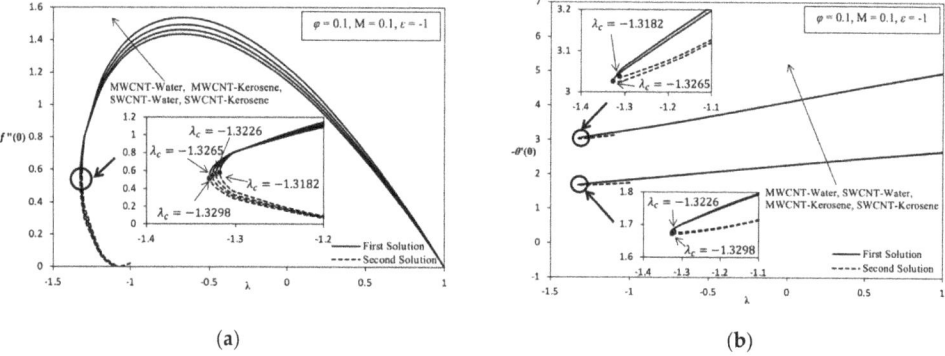

Figure 5. Variation on λ with various values of nanomaterial when $\varphi = 0.1$, $M = 0.1$, and $\varepsilon = -1$: (**a**) Variation on $f''(0)$; (**b**) Variation on $-\theta'(0)$.

Furthermore, Figures 2a, 3a, 5a and 6a described that when $\lambda = 1$ indicates a value of $Re_x^{1/2}C_f$ is zero. This means the velocity fluid flow of stagnation point is equivalent with a velocity wall at rate $\lambda = 1$, which is due to there being no friction that occurs on surface. The point at $(1, 0)$ is also known as the transition point. The second solution shows that the value of $Re_x^{1/2}C_f$ slightly increases when the value of φ increases in momentum with the boundary layer thickness slightly thinning and slightly increasing in skin friction along the shrinking case. The dual solution in Figure 2b shows the reduction in local Nusselt number, $Re_x^{-1/2}Nu_x$ which can be referred to as $-\theta'(0)$, when the value of φ increases from 0 to 0.2. There is a significant reduction in the value of $Re_x^{-1/2}Nu_x$ along the shrinking case because of the existence of the heat sink effect, $\varepsilon < 0$ given that the problems study also considers the parameters, ε to discover various type of behaviors of this model. Consequently, thermal boundary layer thickness δ_T becomes thick when the value of φ increases and the temperature gradient decreases. Thus, this statement proves that when the numerical result with the change value of φ from 0 to 0.2, which is produced in the variation of λ with heat sink effect, is neglected, $\varepsilon = 0$. When this is performed, there is an increase in the value of $Re_x^{-1/2}Nu_x$ along the shrinking case that shows in Figure 6b, and it increases temperature gradient, and the thermal boundary layer thickness becomes thin. However, the value of $Re_x^{1/2}C_f$ continues to decreases, thus increasing the value of φ that is shown in Figure 6a. This proves that nanofluids have a better heat enhancement compared with fluid $\varphi = 0$. Because of the existence of heat sink effect, $\varepsilon < 0$ in this model further inhibits the heat transfer rate of increasing φ. Thus, it is worth noting that if the parameter φ is applied, only a few values of φ used in the base fluid are enough for heat transfer enhancement.

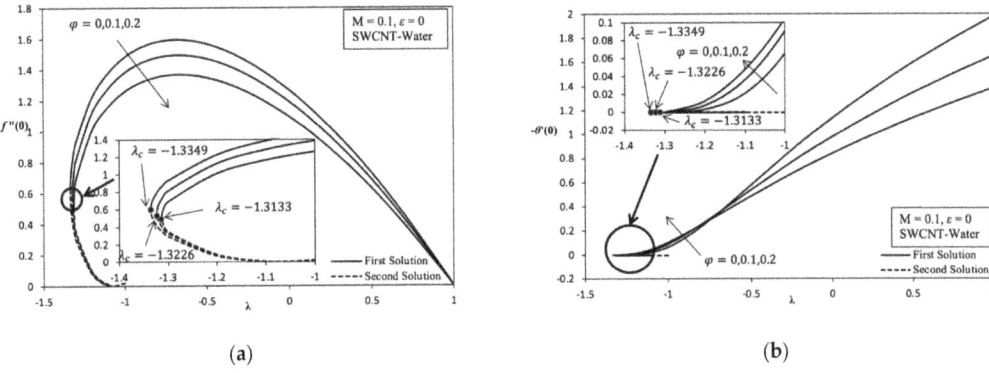

Figure 6. Variation on λ with various values of φ in SWCNT-water when $M = 0.1$: (a) Variation on $f''(0)$; (b) Variation on $-\theta'(0)$.

In addition, Figure 3a also highlights the increasing value of $Re_x^{1/2}C_f$ along with the increasing value of magnetic parameter, M, in the stretching/shrinking case. This is happens because of the Lorentz force $J \times B$, which is equivalent to drag or the viscosity force acting on the surface. However, it is opposed by fluid flow. It can significantly increase the shear stress on the shrinking surface. Thus, the momentum boundary layer thickness decreases with the increasing value of M. The second solution described the decreasing value of $Re_x^{1/2}C_f$ when M increases from 0 to 0.2. The dual solution in Figure 3b shows a slight increase in $Re_x^{-1/2}Nu_x$ when the value of M is higher from 0 to 0.2. Then, the heat transfer rate increases because of the thermal boundary layer thickness becoming thin, and it causes a temperature gradient increase. Thus, the existence of MHD stagnation flow of nanofluid gives a good impact in terms of heat transfer enhancement in application terms, for example, the heat exchanger process and cooling system.

Moreover, it was found that the increasing value of $Re_x^{-1/2}Nu_x$ goes along with increasing value of heat sink parameter ε from 0 to -2 as shown in Figure 4. As far as we know, the value of $Re_x^{1/2}C_f$ does not show any change of parameter ε or, in other words, is uniform because this parameter does not depend on momentum. This indicates that the thermal boundary layer thickness decreases when $\varepsilon < 0$ increases, causing the temperature gradient to be higher. Therefore, heat transfer enhancement is improving with the increasing heat sink effect, which is commonly found in application cooling systems on electronic devices. Figure 5a,b shows the variation of $Re_x^{1/2}C_f$ and $Re_x^{-1/2}Nu_x$, respectively, on the different of nanofluids selected for solving the problem study which are SWCNT-kerosene, SWCNT-water, MWCNT-kerosene, and MWCNT-water. Based on Figure 5a, it is indicated that the value of $Re_x^{1/2}C_f$ of SWCNT-kerosene is the highest compared with the different nanofluids followed by SWCNT-water, MWCNT-kerosene, and MWCNT-water. This is because the SWCNT nanoparticle is better than MWCNT, as well as kerosene being higher than the value of the Prandtl number than water which is, respectively, around 21 and 6.2. Thus, the momentum boundary layer thickness on the different of nanofluid is followed by a thick layer, which is made up of MWCNT-water, MWCNT-kerosene, SWCNT-water, and SWCNT-kerosene. Based on Figure 5b, it was proven that SWCNT-kerosene nanofluid have shown the highest value of $Re_x^{-1/2}Nu_x$

followed by MWCNT-kerosene, SWCNT-water, and MWCNT-water. The MWCNT-water nanofluid is among the most deteriorating nanofluid in terms the value of $Re_x^{1/2}C_f$ and $Re_x^{-1/2}Nu_x$, which cause the momentum and thermal boundary layer thickness to become thick, thus affecting the heat transfer process. Hence, this conclusion is justified based on the velocity and temperature profile produced in Figure 7a,b.

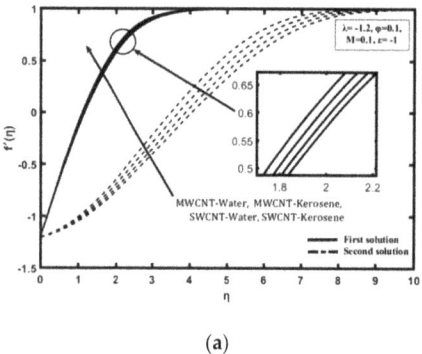
(a)

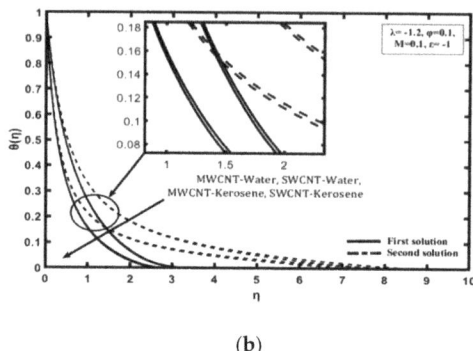
(b)

Figure 7. Various CNT and based fluid when $M = 0.1$, $\lambda = -1.2$, $\varphi = 0.1$, and $\varepsilon = -1$: (a) Velocity profile; (b) Temperature profile.

Overall, it is indicated that the variation of $Re_x^{1/2}C_f$ and $Re_x^{-1/2}Nu_x$ in each parameter, which are φ, M, ε, and the different nanofluids are able to provide their own critical value. Based on Figures 2–6 for the variation result, it is shown that the different value of λ in the more shrinking case causes a significant increase in the value of $Re_x^{1/2}C_f$ while significantly decreasing the value of $Re_x^{-1/2}Nu_x$.

4. Conclusions

This research is about MHD stagnation point flow and heat transfer of nanofluids over a shrinking surface with the heat sink effect being analyzed numerically and discussed in detail in this paper. It was found that the involved parameters such as magnetic parameter, heat sink effect, different types of nanoparticles and base fluids significantly affect the flow and heat transfer. It was found that a decrease in both velocity and temperature was observed with an increase in the volume fraction nanoparticle of the carbon nanotube parameter when the heat sink effect existed, $\varepsilon < 0$. When there is no heat sink effect, $\varepsilon = 0$ for the shrinking case, and the velocity decreases while the temperature increases. Dual solutions exist up to a certain range of the shrinking parameter. This study noticed that both the skin friction coefficient and local Nusselt number increased with an increase in the Magnetic parameter. It was also observed that the values of $f''(0)$ were unchanged while the values of $-\theta'(0)$ increased with the increase in heat sink parameter, $\varepsilon < 0$. Based on the type of nanofluid, SWCNT-kerosene has the highest skin friction coefficient and local Nusselt number. Through the literature conducted for a similar problem, it was concluded that SWCNT-kerosene nanofluid tends toward heat transfer enhancement more than other nanofluid types. Moreover, nanofluid is better than conventional fluid because of local Nusselt number, with $-\theta'(0)$ both increasing more than conventional fluid when the heat sink effect does not exist, $\varepsilon = 0$ and the velocity parameter is near to critical value of λ_c or shrinking case, $\lambda < 0$, although the skin friction coefficient, $f''(0)$ decreases.

Author Contributions: M.N.O., writing—original draft preparation; A.J., supervision, writing—review and editing; N.A.A.B., writing—review and editing. All authors have read and agreed to the published version of the manuscript.

Funding: This research was funded by Universiti Kebangsaan Malaysia, grant No. GGP-2020-030.

Data Availability Statement: Not applicable.

Conflicts of Interest: The authors declare no conflict of interest.

Sample Availability: Not applicable.

Nomenclature

Symbols		Greek Symbols	
c, a	Positive constant	ε	Heat sink/source parameter
B_0	Magnetic field (induced)	σ	Electrical conductivity
C_f	The coefficient of skin friction	μ_{nf}	Nanofluid viscosity
M	Magnetic	$(\rho C_p)_{nf}$	Nanofluid heat capacity
Nu_x	Nusselt number;	ρ	Density
Pr	Prandtl number	λ	Stretching/shrinking parameter
Rd	Thermal radiation	k	Thermal conductivity
Re_x	Reynold number	φ	Volume fraction of CNT
T	Fluid temperature	Subscript	
u, v	Velocity component (x- and y-axes)	nf	Nanofluid
U_w	Velocity (stretching/shrinking sheet)	f	Base fluid
		s	Solid
	Superscript	c	Critical
$(')$	Prime denotes differentiation with respect to η	∞	Far field condition/ambient
		w	Surface condition

References

1. Wang, X.Q.; Mujumdar, A.S. Heat transfer characteristics of nanofluids: A review. *Int. J. Therm. Sci.* **2007**, *46*, 1–19. [CrossRef]
2. Asirvatham, L.G.; Raja, B.; Mohan Lal, D.; Wongwises, S. Convective heat transfer of nanofluids with correlations. *Particuology* **2011**, *9*, 626–631. [CrossRef]
3. Kumar, N.; Singh, P.; Redhewal, A.K.; Bhandari, P. A Review on Nanofluids Applications for Heat Transfer in Micro-channels. *Procedia Eng.* **2015**, *127*, 1197–1202. [CrossRef]
4. Vanaki, S.M.; Ganesan, P.; Mohammed, H.A. Numerical study of convective heat transfer of nanofluids: A review. *Renew. Sustain. Energy Rev.* **2016**, *54*, 1212–1239. [CrossRef]
5. Han, D.; He, W.F.; Asif, F.Z. Experimental study of heat transfer enhancement using nanofluid in double tube heat exchanger. *Energy Procedia* **2017**, *142*, 2547–2553. [CrossRef]
6. Chiam, H.W.; Azmi, W.H.; Adam, N.M.; Ariffin, M.K.A.M. Numerical study of nanofluid heat transfer for different tube geometries—A comprehensive review on performance. *Int. Commun. Heat Mass Transf.* **2017**, *86*, 60–70. [CrossRef]
7. Ahmadi, M.; Willing, G. Heat transfer measurement in water based nanofluids. *Int. J. Heat Mass Transf.* **2018**, *118*, 40–47. [CrossRef]
8. Bowers, J.; Cao, H.; Qiao, G.; Li, Q.; Zhang, G.; Mura, E.; Ding, Y. Flow and heat transfer behaviour of nanofluids in microchannels. *Prog. Nat. Sci. Mater. Int.* **2018**, *28*, 225–234. [CrossRef]
9. Buschmann, M.H.; Azizian, R.; Kempe, T.; Juliá, J.E.; Martínez-Cuenca, R.; Sundén, B.; Wu, Z. Correct interpretation of nanofluid convective heat transfer. *Int. J. Therm. Sci.* **2018**, *129*, 504–531. [CrossRef]
10. Fauzi, N.F.; Ahmad, S.; Pop, I. Stagnation point flow and heat transfer over a nonlinear shrinking sheet with slip effects. *Alex. Eng. J.* **2015**, *54*, 929–934. [CrossRef]
11. Bhatti, M.M.; Shahid, A.; Rashidi, M.M. Numerical simulation of Fluid flow over a shrinking porous sheet by Successive linearization method. *Alex. Eng. J.* **2016**, *55*, 51–56. [CrossRef]
12. Soid, S.K.; Kechil, S.A.; Ishak, A. Axisymmetric stagnation-point flow over a stretching/shrinking plate with second-order velocity slip. *Propuls. Power Res.* **2016**, *5*, 194–201. [CrossRef]
13. Dash, G.C.; Tripathy, R.S.; Rashidi, M.M.; Mishra, S.R. Numerical approach to boundary layer stagnation-point flow past a stretching/shrinking sheet. *J. Mol. Liq.* **2016**, *221*, 860–866. [CrossRef]
14. Hayat, T.; Khan, M.I.; Waqas, M.; Alsaedi, A.; Farooq, M. Numerical simulation for melting heat transfer and radiation effects in stagnation point flow of carbon–Water nanofluid. *Comput. Methods Appl. Mech. Eng.* **2017**, *315*, 1011–1024. [CrossRef]
15. Shit, G.C.; Haldar, R.; Mandal, S. Entropy generation on MHD flow and convective heat transfer in a porous medium of exponentially stretching surface saturated by nanofluids. *Adv. Powder Technol.* **2017**, *28*, 1519–1530. [CrossRef]
16. Uddin, M.S.; Bhattacharyya, K. Thermal boundary layer in stagnation-point flow past a permeable shrinking sheet with variable surface temperature. *Propuls. Power Res.* **2017**, *6*, 186–194. [CrossRef]

17. Abbas, N.; Saleem, S.; Nadeem, S.; Alderremy, A.A.; Khan, A.U. On stagnation point flow of a micro polar nanofluid past a circular cylinder with velocity and thermal slip. *Results Phys.* **2018**, *9*, 1224–1232. [CrossRef]
18. Kumar, R.; Sood, S.; Raju, C.S.K.; Shehzad, S.A. Hydromagnetic unsteady slip stagnation flow of nanofluid with suspension of mixed bio-convection. *Propuls. Power Res.* **2019**, *8*, 362–372. [CrossRef]
19. Mustafa, I.; Abbas, Z.; Arif, A.; Javed, T.; Ghaffari, A. Stability analysis for multiple solutions of boundary layer flow towards a shrinking sheet: Analytical solution by using least square method. *Phys. A Stat. Mech. Its Appl.* **2020**, *540*, 123028. [CrossRef]
20. Al-Amri, F.; Muthtamilselvan, M. Stagnation point flow of nanofluid containing micro-organisms. *Case Stud. Therm. Eng.* **2020**, *21*, 100656. [CrossRef]
21. Anuar, N.S.; Bachok, N.; Arifin, N.M.; Rosali, H. MHD flow past a nonlinear stretching/shrinking sheet in carbon nanotubes: Stability analysis. *Chin. J. Phys.* **2020**, *65*, 436–446. [CrossRef]
22. Tshivhi, K.S.; Makinde, O.D. Magneto-nanofluid coolants past heated shrinking/stretching surfaces: Dual solutions and stability analysis. *Results Eng.* **2021**, *10*, 100229. [CrossRef]
23. Tadesse, F.B.; Makinde, O.D.; Enyadene, L.G. Hydromagnetic stagnation point flow of a magnetite ferrofluid past a convectively heated permeable stretching/shrinking sheet in a Darcy–Forchheimer porous medium. *Sādhanā* **2021**, *46*, 115. [CrossRef]
24. Tadesse, F.B.; Makinde, O.D.; Enyadene, L.G. Mixed Convection of a Radiating Magnetic Nanofluid past a Heated Permeable Stretching/Shrinking Sheet in a Porous Medium. *Math. Probl. Eng.* **2021**, *2021*, 6696748. [CrossRef]
25. Bahiraei, M. Studying nanoparticle distribution in nanofluids considering the effective factors on particle migration and determination of phenomenological constants by Eulerian–Lagrangian simulation. *Adv. Powder Technol.* **2015**, *26*, 802–810. [CrossRef]
26. Liu, Z.; Clausen, J.R.; Rekha, R.R.; Aidun, C.K. A unified analysis of nano-to-microscale particle dispersion in tubular blood flow. *Phys. Fluids* **2019**, *31*, 081903. [CrossRef]
27. Liu, Z.; Zhu, Y.; Clausen, J.R.; Lechman, J.B.; Rao, R.R.; Aidun, C.K. Multiscale method based on coupled lattice-Boltzmann and Langevin-dynamics for direct simulation of nanoscale particle/polymer suspensions in complex flows. *Int. J. Numer. Meth. Fluids* **2019**, *91*, 228–246. [CrossRef]
28. Bachok, N.; Ishak, A.; Pop, I. Stagnation-point flow over a stretching/shrinking sheet in a nanofluid. *Nanoscale Res. Lett.* **2011**, *6*, 623. [CrossRef]
29. Wang, C.Y. Stagnation flow towards a shrinking sheet. *Int. J. Non Linear Mech.* **2008**, *43*, 377–382. [CrossRef]

Article

Computational Investigations of a pH-Induced Structural Transition in a CTAB Solution with Toluic Acid

Tingyi Wang [1], Hui Yan [2,*], Li Lv [3], Yingbiao Xu [1], Lingyu Zhang [1] and Han Jia [4]

[1] Technology Inspection Center, Shengli Oilfield Company, SINOPEC, Dongying 257000, China; wangtingyi180.slyt@sinopec.com (T.W.); xuyingbiao.slyt@sinopec.com (Y.X.); zhangly639.slyt@sinopec.com (L.Z.)
[2] School of Pharmaceutical Sciences, Liaocheng University, Liaocheng 252059, China
[3] Changqing Well Technology Work Company, Chuanqing Drilling Engineering Company Limlted, CNPC, Xi'an 710021, China; cqjxlvli123@163.com
[4] Shandong Key Laboratory of Oilfield Chemistry, School of Petroleum Engineering, China University of Petroleum (East China), Qingdao 266580, China; jiahan@upc.edu.cn
* Correspondence: yanhui@lcu.edu.cn

Citation: Wang, T.; Yan, H.; Lv, L.; Xu, Y.; Zhang, L.; Jia, H. Computational Investigations of a pH-Induced Structural Transition in a CTAB Solution with Toluic Acid. *Molecules* **2021**, *26*, 6978. https://doi.org/10.3390/molecules26226978

Academic Editors: Shiling Yuan and Heng Zhang

Received: 31 October 2021
Accepted: 16 November 2021
Published: 19 November 2021

Publisher's Note: MDPI stays neutral with regard to jurisdictional claims in published maps and institutional affiliations.

Copyright: © 2021 by the authors. Licensee MDPI, Basel, Switzerland. This article is an open access article distributed under the terms and conditions of the Creative Commons Attribution (CC BY) license (https://creativecommons.org/licenses/by/4.0/).

Abstract: In this work, molecular dynamics simulations were performed to study the pH-induced structural transitions for a CTAB/p-toluic acid solution. Spherical and cylindrical micelles were obtained for aqueous surfactants at pH 2 and 7, respectively, which agrees well with the experimental observations. The structural properties of two different micelles were analyzed through the density distributions of components and the molecular orientations of CTA^+ and toluic acid inside the micelles. It was found that the bonding interactions between CTA^+ and toluic in spherical and cylindrical micelles are very different. Almost all the ionized toluic acid (PTA^-) in the solution at pH 7 solubilized into the micelles, and it was located in the CTA^+ headgroups region. Additionally, the bonding between surfactant CTA^+ and PTA^- was very tight due to the electrostatic interactions. The PTA^- that penetrated into the micelles effectively screened the electrostatic repulsion among the cationic headgroups, which is considered to be crucial for maintaining the cylindrical micellar shape. As the pH decreased, the carboxyl groups were protonated. The hydration ability of neutral carboxyl groups weakened, resulting in deeper penetration into the micelles. Meanwhile, their bonding interactions with surfactant headgroups also weakened. Accompanied by the strengthen of electrostatic repulsion among the positive headgroups, the cylindrical micelle was broken into spherical micelles. Our work provided an atomic-level insights into the mechanism of pH-induced structural transitions of a CTAB/p-toluic solution, which is expected to be useful for further understanding the aggregate behavior of mixed cationic surfactants and aromatic acids.

Keywords: molecular dynamics simulation; pH-induced structural transitions; rodlike micelle; sphecial micelle; cationic surfactant

1. Introduction

The controllable self-assemblies of the amphiphilic molecules in aqueous solution are hot issues in both scientific and technological areas [1,2]. The size and shape of the surfactant assembly mainly depend on the chemical structures of the surfactants, such as the lengths of the hydrocarbon chains, properties of the polar headgroups, and the counter ions [3–6]. Generally, surfactants in solution form spherical micelles spontaneously above the critical micelle concentration (CMC) [7–9]. With a further increase in surfactant concentration, the spherical micelles may grow into rod-like or wormlike micelles, and even vesicles.

Adding certain amounts of simple inorganic ions (such as Cl^- and Br^-) or aromatic anions (such as salicylate and benzoic acid) into cationic surfactant solutions can also lead the formation of long rod-like or wormlike micelles at a lower surfactant concentration [9–13].

Due to the stimuli of functional groups in the additive salts, these aggregations consisting of cationic surfactant and anionic additives are sensitive to the external conditions [14–18]. Under an external stimulus, such as pH, temperature, or UV/vis, structural micellar transitions may occur. Thus, controllable self-assemblies of the surfactant in solution can be realized as desired. These stimuli-responsive surfactant systems have attracted much attention in fundamental research and industrial applications, such as drug release, soil remediation, and oilfield industries [1,2].

According to the previous studies on such controllable surfactant systems, it is believed that the interactions between surfactants and the additives are responsive for the stimulus responsiveness of the aggregations. Under external stimuli, no matter what happens to the structure to the surfactant or additive—such as protonated/ionized and cis-transitions—the intermolecular interactions were finally changed. When limited to experimental techniques, it is hard to observe these microscopic interactions directly. Thus, to further investigate the controllable surfactant systems at the molecular level would be significant and useful for understanding the molecular mechanisms behind the specific effects of surfactants or additives on the stimulus-responsiveness performance.

During the past decades, molecular dynamics (MD) simulations have been proven to be a powerful technique to provide supplemental and microscopic insights into experimental observations [19–24]. Many computational studies have been devoted to gain insights into the micro-behavior of the various surfactant systems. However, most previous studies mainly focused on the morphologies of the aggregations. Investigations on the changes in intermolecular interactions inside the aggregations are relatively scarce, especially on the changes in bonding structure induced by external stimuli.

In this work, we studied the structural transitions of a typical cationic surfactant/additive micelle solution induced by pH variation. Cetyltrimethylammonium bromide (CTAB) is one of the most extensively applied cationic surfactants. It forms spherical micelles with a diameter of 2–3 nm when above the CMC in water. These micelles will grow into rod-like or wormlike micelles when the surfactant concentration is far above the CMC (about several hundred times above the CMC). By increasing the ionic strength or adding hydrotrotes into micelle solution, the spherical micelles will undergo a sphere to rod-like shape transition, even at lower concentrations. Besides promoting micellar growth, the aromatic hydrotrotes are sensitive to external conditions, including temperature, UV/vis light, and pH. The aqueous behavior of CTAB in the presence of phenols, salicylate, and aromatic acids has been widely studied [25–29].

The structural transitions of a CTAB/p-toluci acid (PTA) micellar solution were investigated as a representative system in this paper. By altering the pH of a CTAB/p-toluci acid solution, the surfactants can form micelles with different geometries [3]. Our aim was to study the effects of different intermolecular interactions on the structural transitions of CTAB/p-toluci acid aggregations. The simulations started with pre-assembled cylindrical micelles. Experimental observations were successfully reproduced [3]. Based on the MD results, microscopic information on the mechanism behind the pH-induced micellar shape transition has been provided.

2. Results and Discussion

2.1. Different Aggregation Morphologies

Figure 1 shows the simulated configurations of the two systems at the beginning and end of the simulations. As expected, spherical micelles were obtained in the presence of the protonated PTA (pH = 2), whereas a rod-like micelle was obtained when all PTA molecules were deprotonated to PTA$^-$ (pH = 7). From the final configuration, it can be seen that almost all the PTA$^-$ ions were solubilized into the rod-like micelle. In the protonated PTA system, most of the neutral PTA molecules still remained in the water phase. The only two PTA$^-$ ions were solubilized into the micelle. Therefore, it is believed the aggregation shape of the micelle should be related to the quantity of the solubilized additives.

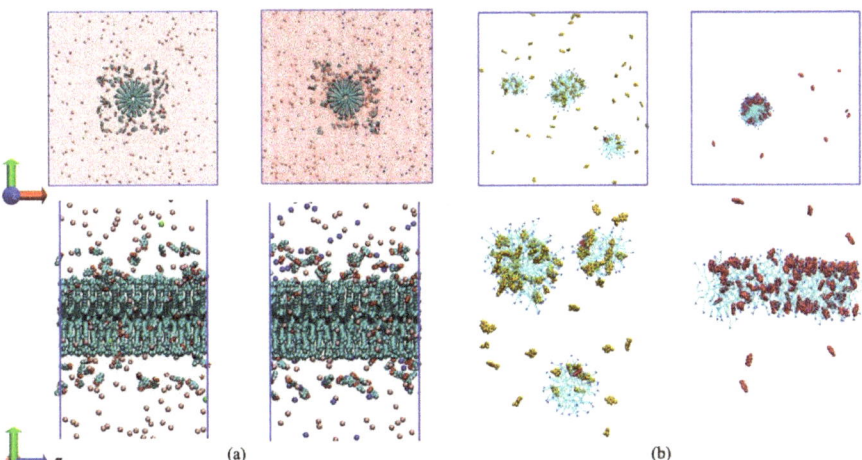

Figure 1. (**a**) Initial setup of the two simulated systems. (**b**) Configurations of CTAB/PTA (pH 2, left in inset a and b) and CTAB/PTA$^-$ (pH 7, right in inset a and b) aggregations. The top panel shows the views from the z-axis perspective. The solid blue lines represent the periodical boundary conditions. Water molecules and inorganic ions are hidden in inset b for clarity. The neutral forms PTA and ionized PTA$^-$ are displayed in yellow and red, respectively.

The absolute number of the solubilized PTA$^-$ into the rod-like micelle was counted as a function of simulation time, as shown in Figure 2. In addition, the radius of the rod-like micelle with time evolution was monitored to show the changes that solubilization brought to the micellar shape. The radius of the rod-like micelle was defined by the average distance between N atoms and the central axis of rod-like micelle. In the initial configuration, the surfactants were loosely packed, yielding a large radius (~2.4 nm) of the pre-assembled micelle. As the simulation went on, the rod-like micelle showed great fluctuation. Meanwhile, the pre-assembled micelle began to shrink due to the hydrophobic interactions between the surfactant chains. A great deal of the PTA$^-$ ions began to enter into the CTA$^+$ aggregation. At about 7.5 ns, the solubilized numbers of PTA$^-$ gradually reached stable values. Subsequently, the fluctuation on the micelle gradually disappeared, resulting in a stable and rigid long rod-like micelle. The radius of the micelle also reached a constant value of about 1.95 nm. The stable aggregated structure indicated the simulation system reached equilibrium, so a total simulation time of 20 ns was sufficient.

2.2. Detailed Structural Properties of the Formed Micelles

As discussed above, the structural transition with the variation in pH is related to the solubilization of the additives into the micelle. Thus, the interactions between additives and surfactants play an important role in stabilizing the micellar structure. Before discussing the intermolecular interactions between additives and surfactants, we must first investigate the distribution of these hydrotropes inside the micelle. The locations of some selected species were characterized by calculating the number density distribution profiles. In Figure 3, the number density distributions were plotted with respect to the central axis of the rod-like micelle, which is along the z-axis of the simulation box. For the spherical micelle, the number density was calculated with respect to the center of mass (COM) of the spherical micelle, i.e., along the radial direction of the spherical micelle. In the simulated system with protonated PTA (pH = 2), three spherical micelles were obtained at the end of the simulation, as shown in Figure 1. We selected the biggest one to calculate the structural properties.

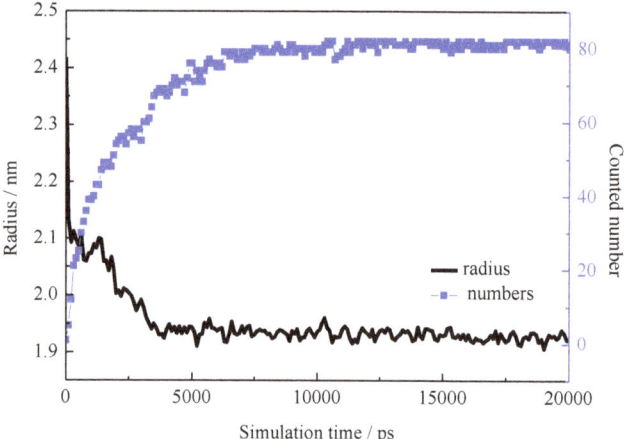

Figure 2. Radius of the rod-like micelle and solubilized numbers of PTA$^-$ in the micelle plotted as a function of simulation time.

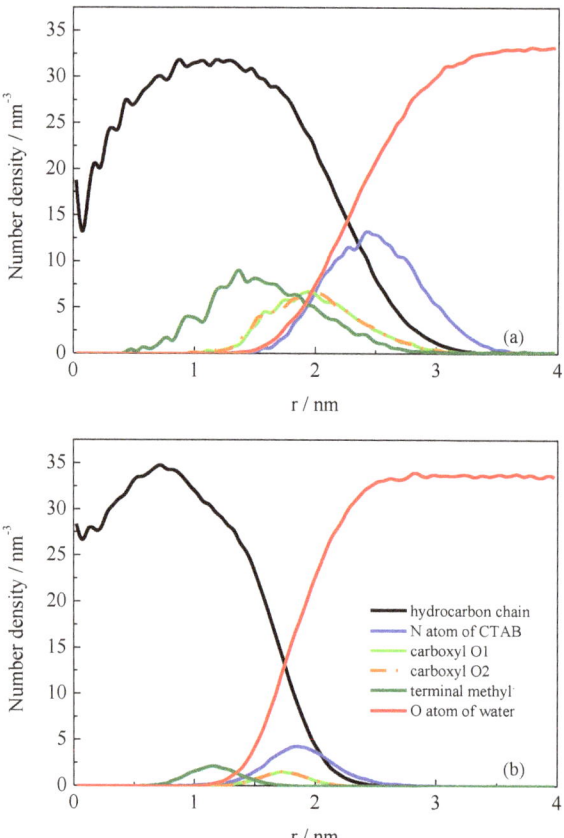

Figure 3. (**a**) Number density distributions of components with respect to the COM of the spherical micelle. Values for N1, O1, O2, and C8 (see Figure 8) were increased 10 times for clarity. (**b**) Number density distributions of components with respect to the central axis of the rod-like micelle.

The results for both the rod-like and spherical micelles are very similar to those of the previous simulation studies [30,31]. It can be seen that the surfactant headgroups which were presented by headgroup N atoms constituted a shell region the surface of the micelle, and the hydrophobic chains were concentrated in the interior in both rod-like and spherical micelles. Our focus is the distribution of the additive molecules PTA or PTA^- inside the micelles. It was found that the terminal methyl groups of PTA or PTA^- were located deeply in the hydrophobic region in both rod-like and special micelles. The carboxyl groups were located on the outer shells of the micelles, and they were adjacent to the surfactant headgroups. This was certainly because that the phenyl groups were hydrophobic and the carboxyl groups were hydrophilic. There are mainly two differences between PTA and PTA^-. One is that the distributions of the two carboxyl O atoms in deprotonated PTA^- ions overlapped, suggesting they were distributed at the same locations inside the rod-like micelle. The peaks in the distributions of the two carboxyl O atoms in protonated PTA^- molecules are staggered. The protonated O2 atoms were located outside a little bit more than the other O1 atoms. The other difference is that the distance between COO^- O atoms and headgroup N atoms was quite short (~0.1 nm), which was measured by the two distribution peaks shown in Figure 3, whereas the distances between COOH O atoms and headgroups N atoms were rather long (~0.7 nm).

The above results show that once the carboxyl groups were protonated with a decrease in pH, the PTA molecules localized more deeply inside the interior of the spherical micelle. This suggests that when the carboxyl groups are changed to be electroneutral, the hydrophobic interactions between methylbenzene groups and CTA^+ hydrocarbon chains will ultimately dominate. The O2 atom in COOH group being located outside a little bit more was mainly due to the stronger interactions between hydroxy groups and water molecules, whereas in the rod-like micelle, the surfactant headgroups were close to the COO^- groups, suggesting strong intermolecular interactions through electrostatic interactions. It is believed that the tight bounding between ammonium groups and COO^- groups plays an important role in maintaining the cylinder micellar shape.

2.3. Bonding Structures of PTA^-/PTA and Surfactants

The detailed interactions between the additive molecules with the surfactants were further investigated by exploring the orientations of PTA^-/PTA inside the micelle. The orientation was defined by the angle θ between the molecular axis of PTA^-/PTA and CTA^+. The molecular axes of PTA^-/PTA and CTA^+ were defined by the vector C8 to C1 (atoms in PTA) and the vector C3 to N (atoms in CTA^+). When calculating the angle, only the neighboring pairs of PTA^-/PTA and CTA^+ molecules were considered; i.e., only the interactive pairs which were judged by their separation distances were counted.

The probability distributions of the angles for PTA^- and PTA are shown in Figure 4. It is evident that the molecular axis of ionized PTA^- preferred to form an angle of about 20° with its adjacent surfactant molecules. When the ionized PTA^- ions were protonated, the distribution of angle between the same vectors became very broad. It can be seen that the value of angle varied from 20° to 90°, suggesting the protonated PTA molecules did not prefer to form certain angles with the surfactants. Figure 4 shows the selected bonding structures between PTA^-/PTA and CTA^+. Obviously, the ionic PTA^- interacted with neighboring CTA^+ through electrostatic interactions between their carboxyl and ammonium groups. The strong electrostatic interactions resulted in tight bonding between PTA^- and CTA^+. While the PTA^- ions were protonated, the strong electrostatic interactions with CTA^+ surfactants disappeared. Therefore, the bonding between surfactants and additives also weakened inside the aggregates, which is considered to be essential for the shaper transition of the micelle.

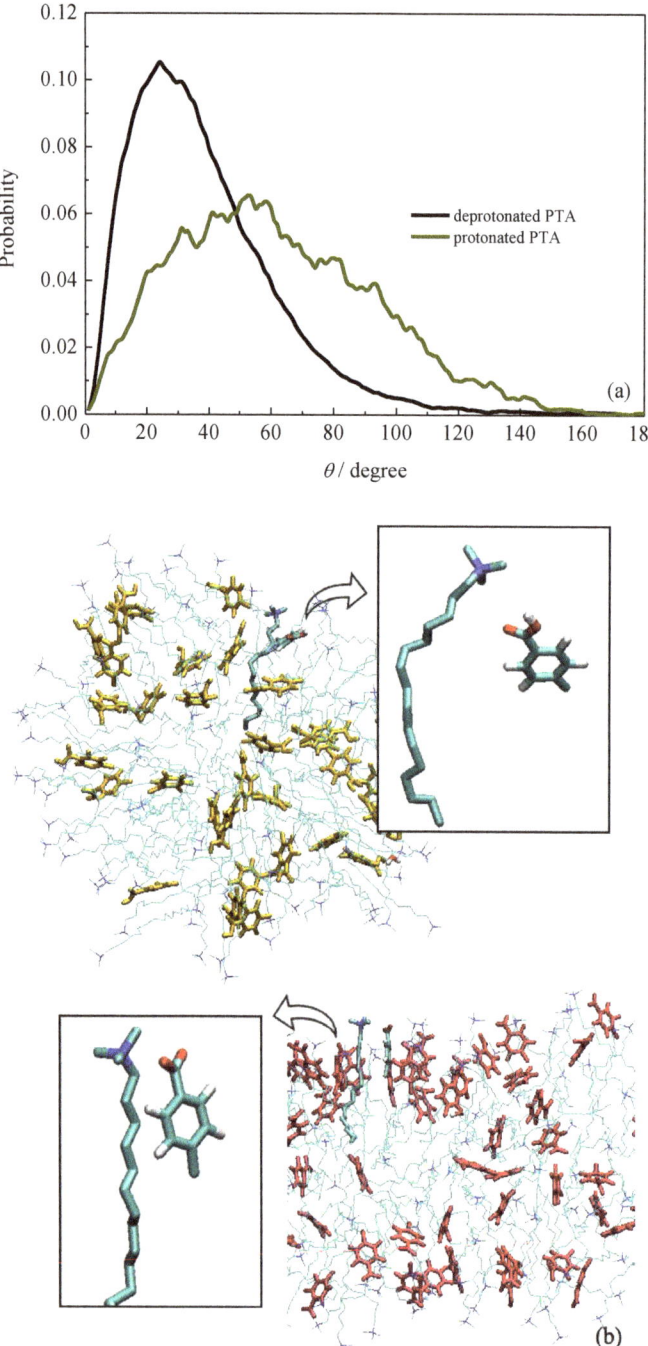

Figure 4. (a) Probability distribution of the angle between vectors defined in the molecular structures. (b) Bonding structures between CTA$^+$ and PTA in spherical and rod-like micelles.

2.4. Intermolecular Interactions

As discussed above, the bonding mode between surfactants and additives may have an influence on the micellar shape. In addition, the surrounding water solution environment may also affect the interior intermolecular interactions. In what follows, some special intermolecular interactions in two micellar systems were investigated to explore the micromechanism behind the micellar shape transition induced by pH variation.

First, the intermolecular interactions between PTA/PTA$^-$ and CTA$^+$ were visualized by analyzing the weak interactions using the Multiwfn software [32]. The reduced density gradient (RDG) was plotted as a function of electron density $\rho(r)$ based on the selected configurations. The gradient isosurfaces were then visualized with the VMD software [32] to show representations of the weak interactions. As shown in Figure 5a, distributions colored in dark blue present interactions between an ionic CTA$^+$ headgroup and PTA$^-$, which correspond to the strong attractive interactions. The attractive interactions were mainly attributed to the electrostatic attraction, wheres, the interaction region between CTA$^+$ and the neutral PTA disappeared. Instead, weak hydrogen bonds may exist between carboxy group and hydrogen atoms in CTA surfactant.

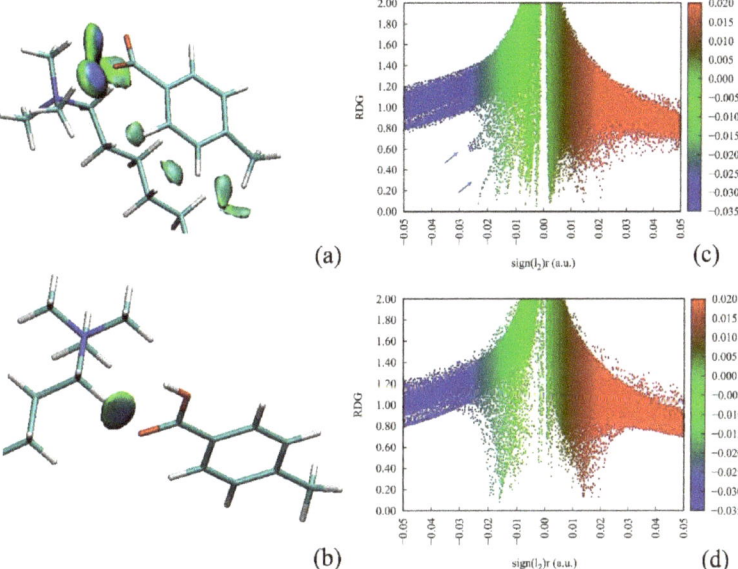

Figure 5. Weak interaction analysis for (**a**) PTA$^-$/CTA$^+$ and (**b**) PTA/CTA$^+$. Insets (**c**,**d**) show the reduced density gradient (RDG) versus electron density for configurations shown in insets a and b, respectively.

The hydration effect of surfactants and additive PTA/PTA$^-$ was then investigated through the radial distribution functions (RDFs). Figure 6a shows the RDFs of water molecules around the carboxyl groups in PTA or PTA$^-$. As shown in RDF profiles, we can see that there were two well-defined hydration shells around the PTA$^-$ carboxyl groups, suggesting ordered arrangement of water molecules around carboxyl groups. The high intensity of the first peak demonstrates strong interactions between the ionized carboxyl groups and water molecules. This kind of interaction fell off rapidly when the ionized carboxyl groups were protonated. Therefore, the oxygen atoms in carboxyl groups of natural PTA molecules were located deeper inside the micelle, as shown in Figure 3.

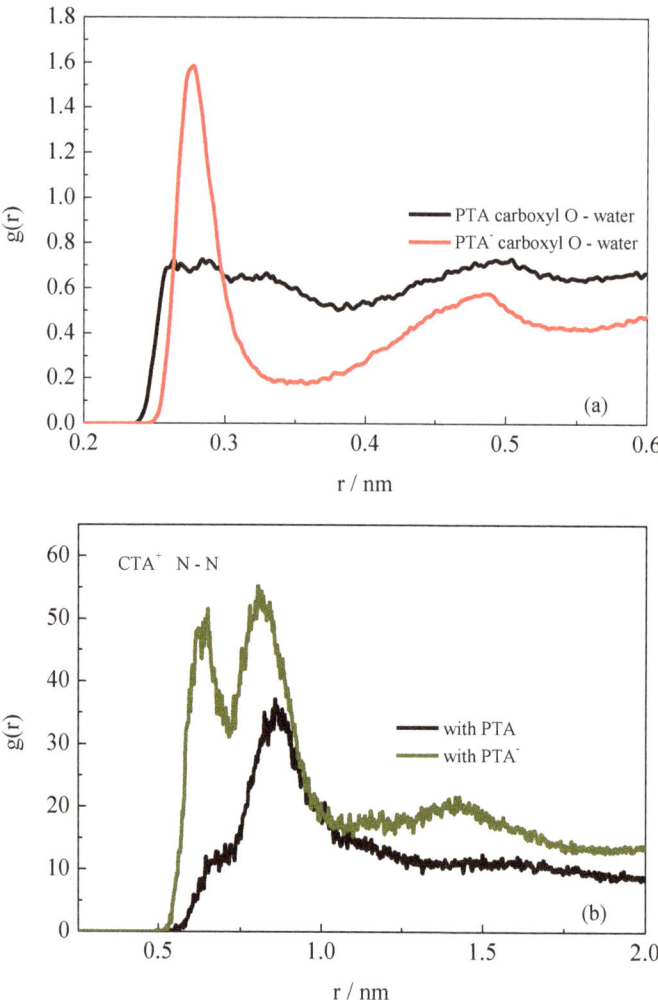

Figure 6. (a) RDFs between surfactant N atoms and water O atoms. (b) RDFs between surfactant N atoms.

Figure 6b shows the RDFs between surfactant CTA^+ headgroup N atoms, which can be used to reflect the aggregating degree among the surfactant headgroups. It can be seen that there were two evident aggregated peaks around surfactant headgroups in the rod-like micelle with the ionized PTA^-. The first peak at about 0.6 nm in its RDF represents the nearest headgroups around one central CTA^+ headgroup, and the second peak at about 0.8 nm represents the headgroups located at the outer shell. In Figure 7, the aggregated structure of the surfactant headgroups in the rod-like micelles is highlighted to show the detailed information. Due to the tight bonding between CTA^+ and PTA^- through electrostatic interactions, the electrostatic repulsion among the positive headgroups was effectively weakened. The electrostatic shielding among the headgroups introduced by PTA^- is therefore considered to play an essential role in maintaining the structure of the rod-like micelle.

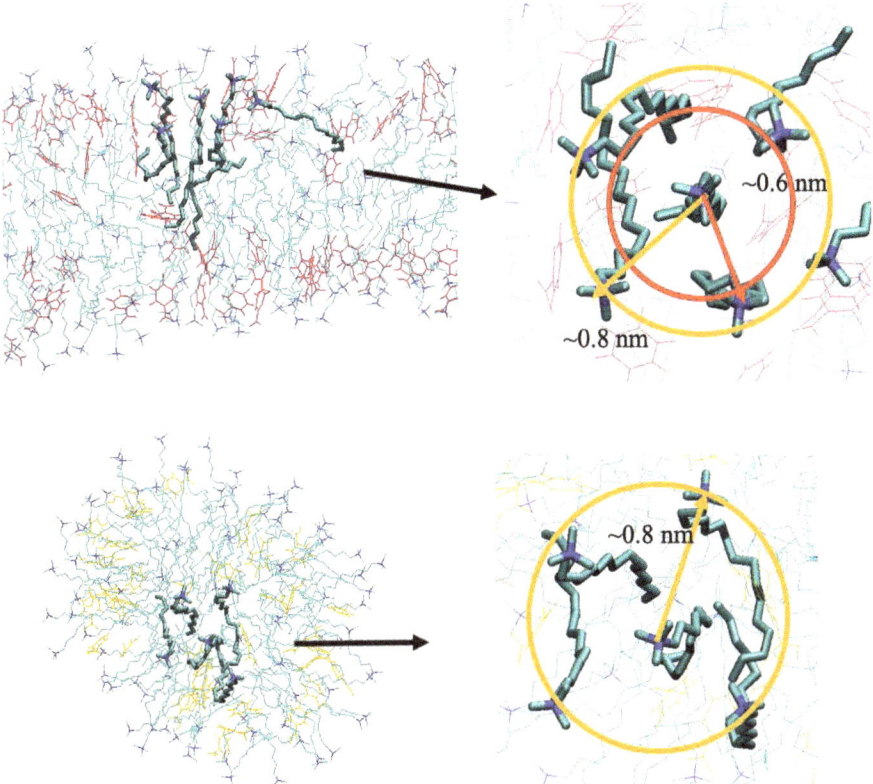

Figure 7. Packed structures of surfactants inside rod-like (**top panel**) and spherical (**bottom panel**) micelles.

For the spherical micelle in the presence of PTA, it is evident that the interactions among the surfactant headgroups weakened greatly. As can be seen from Figure 6b, there was only a shoulder peak at about 0.6 nm. This suggests that the surfactant headgroups were loosely packed, compared with those in the rod-like micelle (Figure 7). Due to the disappearance of the electrostatic shielding from the additive molecules, the positive CTA^+ headgroups repelled each other. Cooperating with the hydrophobic interactions from the surfactant tails, the aggregations prefer to form spherical micelles.

3. Computational Details

First, according to the previous studies [23,24,30,31], a pre-assembled cylindrical micelle was built. The obtained cylindrical micelle consisting of 180 CTA^+ surfactants was placed in a simulation with dimensions of 25 nm × 25 nm × 10 nm. The central axis of the cylindrical micelle was placed centrally in the box along the z direction of the simulation box. Based on the experimental conditions [3], two systems were simulated to investigate the micellar shape transitions induced by pH. The first system was constructed by inserting 90 PTA^- molecules around the pre-assembled micelle, to study the micro-behavior of a CTAB/PTA^- solution at pH 7. The second system corresponded to the situation at pH 2. The acidic environment was represented by adding certain amounts of hydronium and chloride ions. As usual, the hydronium ions were in their hydrated ion forms (H_3O^+). In the acidic situation, 88 PTA^- ions were protonated according to the pKa value of benzoic acid at 298 K. Finally, bromide ions were inserted into the above two systems and the simulation boxes were filled with water molecules. The compositions of two systems are summarized in Table 1.

Table 1. Simulated systems: numbers of each component in the different systems.

Scheme 2.	CTA$^+$	Br$^-$	PTA	PTA$^-$	H$^+$	Cl$^-$	Na$^+$	Water
pH 2	180	180	88	2	9	7		193623
pH 7	180	180		90			90	193636

Molecular dynamics simulations were performed using the Gromacs package (version 2019.3) [33–36]. The united-atom GROMOS 54A7 force field [37] was used to describe the intermolecular interactions. Structures of the surfactant and additives are shown in Figure 8. The force filed parameters for the molecules, including CTA$^+$, PTA/PTA$^-$, and H$_3$O$^+$, were obtained using the Automated Topology Builder (ATB) server [38]. Water molecules were described by the simple point charge/extend (SPC/E) model [39]. The two systems were first minimized through the steepest descent method. Then, a 20 ns MD simulation under the NPT ensemble was performed for each system. During the simulation, the temperature (298 K) and pressure (1 atm) were maintained by the V-rescale thermostat and Berendsen barostat with coupling time constants of 01. and 1.0 ps, respectively [40,41]. LINCS algorithm [42] was applied to constrain the bond lengths of other components. Periodic boundary conditions were applied in all three directions. The cut-off distance for the Lennard–Jones and electrostatic interactions was 1.2 nm. The particle mesh Ewald method was used to calculate the long-range electrostatic interactions [43]. Configurations were visualized using Visual Molecular Dynamics software [44].

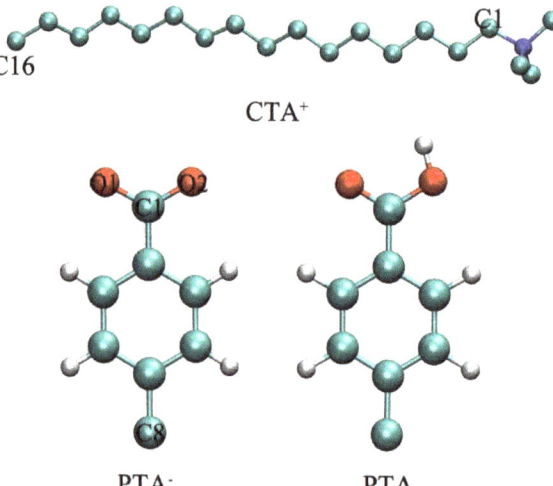

Figure 8. Structures of CTA$^+$, PTA$^-$, and PTA. (Atomic color scheme: C, cyan; O, red; N, blue; and H, white).

4. Conclusions

Molecular dynamics (MD) simulations were performed to investigate the pH-induced structural transitions in aqueous CTAB/PTA solutions. Two simulated systems were created. One was a system consisting of CTAB and neutral PTA, which represented the solution in an acidic environment (pH = 2). The other system consisted of CATB and ionized PTA$^-$ (pH = 7). The two systems were both simulated using a pre-assembled cylindrical micelle. The MD results reproduced the experimental phenomenon—that is, spherical and rod-like micelles were obtained for the systems at pH 2 and 7, respectively.

The mechanism behind the pH-induced micellar shape transitions was investigated on the basis of the MD results. It was found that the ionized PTA^- can effectively screen the electrostatic repulsions among the positive surfactant headgroups, through the strong interactions with surfactant headgroups. The dense packing of the surfactant headgroups lead the formation of a rod-like micelle. With the lower pH, the ionized carboxyl groups were protonated. The bonding of the neutral PTA with surfactant weakened, resulting in the strengthening of electrostatic repulsion among surfactant headgroups. The loose packing among surfactant headgroups resulted in breaking of the cylindrical micelle and the formation of the spherical micelles. Our study provided a molecular mechanism for the pH-induced shape transition in a mixed cationic surfactant and aromatic ions solution. The results presented intuitionistic intermolecular interactions which were responsible for the micellar shape transition. These observations are expected to be useful for the environmental stimuli-responsive colloid systems in experimental studies.

Author Contributions: Data curation, L.L. and Y.X.; Funding acquisition, H.J.; Investigation, T.W. and Y.X.; Project administration, H.Y. and H.J.; Visualization, L.Z.; Writing—original draft, T.W.; Writing—review & editing, H.Y. All authors have read and agreed to the published version of the manuscript.

Funding: This work was funded by the Fundamental Research Funds for the Central Universities (19CX02064A and 19CX05006A), and the Opening Fund of Shandong Key Laboratory of Oilfield Chemistry.

Institutional Review Board Statement: Not applicable.

Informed Consent Statement: Not applicable.

Data Availability Statement: The data presented in this study are available on request from the corresponding author.

Conflicts of Interest: The authors declare no conflict of interest.

Sample Availability: Samples of the compounds are not available from the authors.

References

1. Chu, Z.; Dreiss, C.A.; Feng, Y. Smart wormlike micelles. *Chem. Soc. Rev.* **2013**, *42*, 7174–7203. [CrossRef]
2. Dreiss, C.A. Wormlike micelles: Where do we stand? Recent developments, linear rheology and scattering techniques. *Soft Matter* **2013**, *3*, 956–970. [CrossRef] [PubMed]
3. Patel, V.; Dharaiya, N.; Ray, D.; Aswal, V.K.; Bahadur, P. pH controlled size/shape in CTAB micelles with solubilized polar additives: A viscometry, scattering and spectral evaluation. *Colloid Surf. A* **2014**, *455*, 67–75. [CrossRef]
4. De, S.; Aswal, V.K.; Goyal, P.S.; Bhattacharya, S. Role of spacer chain length in dimeric micellar organization. Small angle neutron scattering and fluorescence studies. *J. Phys. Chem.* **1996**, *100*, 11664–11671. [CrossRef]
5. Varade, D.; Joshi, T.; Aswal, V.K.; Goyal, P.S.; Hassan, P.A.; Bahadur, P. Effect of salt on the micelles of cetyl pyridinium chloride. *Colloids Surf. A Physicochem. Eng. Asp.* **2005**, *259*, 95–101. [CrossRef]
6. Srinivasan, V.; Blankschtein, D. Effect of counterion binding on micellar solution behavior: 1. Molecular-thermodynamic theory of micellization of ionic surfactants. *Langmuir* **2003**, *19*, 9932–9945. [CrossRef]
7. Moulik, S.P.; Haque, M.E.; Jana, P.K.; Das, A.R. Micellar properties of cationic surfactants in pure and mixed states. *J. Phys. Chem.* **1996**, *100*, 701–708. [CrossRef]
8. Giongo, C.V.; Bales, B.L. Estimate of the ionization degree of ionic micelles based on krafft temperature measurements. *J. Phys. Chem. B* **2003**, *107*, 5398–5403. [CrossRef]
9. Lin, Z.; Cai, J.J.; Sciven, L.E.; Davis, H.T. Spherical-to-wormlike micelle transition in CTAB solution. *J. Phys. Chem.* **1994**, *98*, 5984–5993. [CrossRef]
10. Imae, T.; Ikeda, S. Characteristics of rodlike micelles of cetyltrimethylammonium chloride in aqueous NaCl solutions: Their flexibility and the scaling laws in dilute and semidilute regimes. *Colloid Polym. Sci.* **1987**, *265*, 1090–1098. [CrossRef]
11. Raghavan, S.R.; Kaler, E.W. Highly viscoelastic wormlike micellar solutions formed by cationic surfactants with long unsaturated tails. *Langmuir* **2001**, *17*, 300–306. [CrossRef]
12. Hartmann, V.; Cressely, R. Simple salts effects on the characteristics of the shear thickening exhibited by an aqueous micellar solution of CTAB/NaSal. *Europhys. Lett.* **1997**, *40*, 691–696. [CrossRef]
13. Rao, U.R.K.; Manohar, C.; Valaulikar, B.S.; Lyer, R.M. Micellar chain model for the origin of the viscoelasticity in dilute surfactant solutions. *J. Phys. Chem.* **1987**, *91*, 3286–3291. [CrossRef]

14. Sakai, H.; Matsumura, A.; Yokoyama, S.; Saji, T.; Abe, M. Photochemical Switching of Vesicle Formation Using an Azobenzene-Modified Surfactant. *J. Phys. Chem. B* **1999**, *103*, 10737–10740. [CrossRef]
15. Chu, Z.; Feng, Y. Thermo-switchable surfactant gel. *Chem. Commun.* **2011**, *47*, 7191–7193. [CrossRef] [PubMed]
16. Dexter, A.F.; Malcolm, A.S.; Middelberg, A.P.J. Reversible active switching of the mechanical properties of a peptide film at a fluid-fluid interface. *Nat. Mater.* **2006**, *5*, 502–506. [CrossRef] [PubMed]
17. Minkenberg, C.B.; Florusse, L.; Eelkema, R.; Koper, G.J.M.; van Esch, J.H. Triggered self-assembly of simple dynamic covalent surfactants. *J. Am. Chem. Soc.* **2009**, *131*, 11274–11275. [CrossRef] [PubMed]
18. Jessop, P.G.; Mercer, S.M.; Heldebrant, D.J. CO_2-triggered switchable solvents, surfactants, and other materials. *Energy Env. Sci.* **2012**, *5*, 7240–7253. [CrossRef]
19. Sambasivam, A.; Dhakal, S.; Sureshkumar, R. Structure and rheology of self-assembled aqueous suspensions of nanoparticles and wormlike micelles. *Mol. Simul.* **2018**, *44*, 485–493. [CrossRef]
20. Maillet, J.B.; Lachet, V.; Coveney, P.V. Large scale molecular dynamics simulations of self-assembly processes in short and long chain cationic surfactants. *Phys. Chem. Chem. Phys.* **1999**, *1*, 5277–5290. [CrossRef]
21. Marrink, S.J.; Tieleman, D.P.; Mark, A.E. Molecular dynamics simulation of the kinetics of spontaneous micelle formation. *J. Phys. Chem. B* **2000**, *104*, 12165–12173. [CrossRef]
22. Yakovlev, D.S.; Boek, E.S. Molecular dynamics simulations of mixed cationic/anionic wormlike micelles. *Langmuir* **2007**, *23*, 6588–6597. [CrossRef] [PubMed]
23. Wang, Z.W.; Larson, R.G. Molecular dynamics simulations of threadlike cetyltrimethylammonium chloride micelles: Effect of sodium chloride and sodium salicylate salts. *J. Phys. Chem. B* **2009**, *113*, 13697–13710. [CrossRef] [PubMed]
24. Lorenz, C.D.; Hsieh, C.M.; Dreiss, C.A.; Lawrence, M.J. Molecular dynamics simulations of the interfacial and structural properties of dimethyldodecylamine-N-oxide micelles. *Langmuir* **2011**, *27*, 546–553. [CrossRef]
25. Aswal, V.K.; Goyal, P.S. Role of counterion distribution on the structure of micelles in aqueous salt solutions: Small-angle neutron scattering study. *Chem. Phys. Lett.* **2002**, *357*, 491–497. [CrossRef]
26. Brinchi, L.; Germani, R.; Di Profio, P.; Marte, L.; Savelli, G.; Oda, R.; Berti, D. Viscoelastic solutions formed by worm-like micelles of amine oxide surfactant. *J. Colloid Interface Sci.* **2010**, *346*, 100–106. [CrossRef]
27. Kumar, S.; Sharma, D.; Kabir-ud-Din. Role of partitioning site in producing vis-coelasticity in micellar solutions. *J. Surf. Deterg.* **2005**, *8*, 247–252. [CrossRef]
28. Sabatino, P.; Szczygiel, A.; Sinnaeve, D.; Hakimhashemi, M.; Saveyn, H.; Martins, J.C.; Meeren, P.V. NMR study of the influence of pH on phenol sorption in cationic CTAB micellar solutions. *Colloids Surf. A* **2010**, *370*, 42–48. [CrossRef]
29. Umeasiegbu, C.D.; Balakotaiah, V.; Krishnamoorti, R. pH-Induced re-entrant microstructural transitions in cationic surfactant-hydrotrope mixtures. *Langmuir* **2016**, *32*, 655–663. [CrossRef]
30. Yan, H.; Wang, Y.; Zhang, L.; Li, G.; Wei, X.; Liu, C. Molecular dynamics simulation of spherical-to-threadlike transition in a cationic surfactant solution. *Mol. Simulat.* **2019**, *45*, 797–805. [CrossRef]
31. Yan, H.; Han, Z.; Li, K.; Li, G.; Wei, X. Molecular dynamics simulation of the pH-induced structural transitions in CTAB/NaSal solution. *Langmuir* **2018**, *34*, 351–358. [CrossRef] [PubMed]
32. Lu, T.; Chen, F. Multiwfn: A multifunctional wavefunction analyzer. *J. Comput. Chem.* **2012**, *33*, 580–592. [CrossRef]
33. Abraham, M.J.; Murtola, T.; Schulz, R.; Páll, S.; Smith, J.C.; Hess, B.; Lindahl, E. GROMACS: High performance molecular simulations through multi-level parallelism from laptops to supercomputers. *SoftwareX* **2015**, *1*, 19–25. [CrossRef]
34. Páll, S.; Abraham, M.J.; Kutzner, C.; Hess, B.; Lindahl, E. Tackling exascale software challenges in molecular dynamics simulations with GROMACS. *Solving Softw. Chall. Exascale* **2015**, *8759*, 3–27.
35. Pronk, S.; Páll, S.; Schulz, R.; Larsson, P.; Bjelkmar, P.; Apostolov, R.; Shirts, M.R.; Smith, J.C.; Kasson, P.M.; van der Spoel, D.; et al. GROMACS 4.5: A high-throughput and highly parallel open source molecular simulation toolkit. *Bioinformatics* **2013**, *29*, 845–854. [CrossRef]
36. Hess, B.; Kutzner, C.; van der Spoel, D.; Lindahl, E. GROMACS 4: Algorithms for highly efficient load-balanced, and scalable molecular simulation. *J. Chem. Theory Comput.* **2008**, *4*, 435–447. [CrossRef]
37. Huang, W.; Lin, Z.; van Gunsteren, W.F. Validation of the GROMOS 54A7 force field with respect to beta-peptide folding. *J. Chem. Theory Comput.* **2011**, *7*, 1237–1243. [CrossRef] [PubMed]
38. Malde, A.K.; Zuo, L.; Breeze, M.; Stroet, M.; Poger, D.; Nair, P.C.; Oostenbrink, C.; Mark, A.E. An automated force field topology builder (ATB) and repository: Version 1.0. *J. Chem. Theory Comput.* **2011**, *7*, 4026–4037. [CrossRef]
39. Berendsen, H.J.C.; Grigera, J.R.; Straatsma, T.P. The missing term in effective pair potentials. *J. Phys. Chem.* **1987**, *91*, 6269–6271. [CrossRef]
40. Bussi, G.; Donadio, D.; Parrinello, M. Canonical sampling through velocity rescaling. *J. Chem. Phys.* **2007**, *126*, 014101. [CrossRef]
41. Berendsen, H.J.C.; Postma, J.P.M.; van Gunsteren, W.F.; DiNola, A.; Haak, J.R. Molecular dynamics with coupling to an external bath. *J. Chem. Phys.* **1984**, *81*, 3684–3690. [CrossRef]
42. Hess, B.; Bekker, H.; Berendsen, H.J.C.; Fraaije, J.G.E.M. LINCS: A linear constraint solver for molecular simulations. *J. Comput. Chem.* **1997**, *18*, 1463–1472. [CrossRef]
43. Essman, U.; Perera, L.; Berkowitz, M.L.; Darden, T.; Lee, H.; Pedersen, L.G. A smooth particle-mesh ewald method. *J. Chem. Phys.* **1995**, *103*, 8577–8593. [CrossRef]
44. Humphrey, W.; Dalke, A.; Schulten, K. VMD: Visual molecular dynamics. *J. Mol. Graphics.* **1996**, *14*, 33–38. [CrossRef]

Article

Adsorption of Mussel Protein on Polymer Antifouling Membranes: A Molecular Dynamics Study

Fengfeng Gao

Department of Chemical Engineering, Zibo Vocational Institute, Zibo 255300, China; 11364@zbvc.edu.cn; Tel.: +86-533-2348221

Abstract: Biofouling is one of the most difficult problems in the field of marine engineering. In this work, molecular dynamics simulation was used to study the adsorption process of mussel protein on the surface of two antifouling films—hydrophilic film and hydrophobic film—trying to reveal the mechanism of protein adsorption and the antifouling mechanism of materials at the molecular level. The simulated conclusion is helpful to design and find new antifouling coatings for the experiments in the future.

Keywords: antifouling membrane; protein adsorption; mussel protein; hydrophilic film; hydrophobic film

Citation: Gao, F. Adsorption of Mussel Protein on Polymer Antifouling Membranes: A Molecular Dynamics Study. *Molecules* **2021**, *26*, 5660. https://doi.org/10.3390/molecules26185660

Academic Editors: Shiling Yuan and Heng Zhang

Received: 17 August 2021
Accepted: 14 September 2021
Published: 17 September 2021

Publisher's Note: MDPI stays neutral with regard to jurisdictional claims in published maps and institutional affiliations.

Copyright: © 2021 by the author. Licensee MDPI, Basel, Switzerland. This article is an open access article distributed under the terms and conditions of the Creative Commons Attribution (CC BY) license (https://creativecommons.org/licenses/by/4.0/).

1. Introduction

Marine fouling is one common problem for ships and marine facilities. The adhesion of marine organisms leads to the increase of hull surface roughness, which can increase the resistance of navigation, and the organic acids released by these organisms also accelerates the corrosion of ships and marine facilities. Worldwide, the cost of fuel consumption, hull cleaning, painting, and maintenance caused by marine fouling is about several billions of dollars per year.

In order to reduce the loss caused by marine fouling, the common method is to apply antifouling coating on the surface of ship. The traditional antifouling coating is mixed with organic copper, organic tin, etc., and these coating are mainly to kill the marine organisms attached to the hull surface by releasing heavy metal ions. However, these heavy metal ions accumulate in the food chain and eventually endanger human beings [1]. At present, designing and developing new environmentally friendly marine antifouling coatings has become a hot new trend in the field of biological antifouling.

There are many kinds of adhesion marine organisms in the ocean, including algae, shellfish, sponge, etc. Among them, shellfish (such as barnacles, mussels, etc.) are more difficult to treat. Normally, the attachment mechanism of them depends on their own released proteins temporarily or permanently adhering to the solid surface (such as the hull). At present, the adhesion materials being widely studied are barnacle glue and mussel protein in most of research experiments. In these experiments, the adsorption of proteins on the surface of materials is related to their own properties (such as surface charge, hydrophobicity, conformation, etc. [2,3]) and the materials surface's properties (such as surface roughness, chemical composition, etc.).

Till now, there has not been a unanimous view about antifouling mechanism. However, it is generally known that the hydration layer on the surface of materials has a great relationship with its antifouling effect [4–6]. A large number of experimental studies show that the antifouling effect of a hydrophilic surface is better than that of a hydrophobic surface [7,8]. This contributed that one closely bonded hydration layer can be formed on the surface of hydrophilic materials, and the adsorption process of proteins must destroy the hydration layer firstly. Therefore, the binding ability between water molecules and membrane materials can reflect the difficulty of binding between proteins and solid surfaces [9–12].

It is very important to understand the interaction mechanism between protein and solid surface for both theoretical and applied research. In order to design and find new antifouling materials, it is necessary to study the adsorption mechanism between biological proteins and antifouling materials. However, it is very difficult to reveal the internal mechanism of adsorption at the molecular level for any experimental technique. In this paper, we used molecular dynamics method to study the adsorption behavior of mussel protein on different materials surface, and we will try to explain the adsorption mechanism of mussel protein and the antifouling mechanism of the material surface on the microscopic level.

2. Simulation Method

Polymer antifouling membrane is constructed by Amorphous cell module in Materials Studio Software (version 4.4). Firstly, the dimethylsiloxane (PDMS) layer was constructed. Its density is 0.965 g/cm^3 and the thickness of the membrane is about 2 nm. Then, alkyl chains of -(CH$_2$)$_{10}$CH$_3$ and -(CH$_2$)$_{10}$COOH are grafted on the surface of PDMS, respectively, and the grafting number density is about 0.66 per square nanometer. Finally, CH$_3$-SAM and COOH-SAM polymer antifouling membrane are obtained. Both represent the hydrophobic and hydrophilic antifouling membrane, respectively [13–16].

In the simulation, mussel protein (PDB ID: 5DUY) is used as the model protein, which contains 150 amino acid molecules. It is similar to the spherical structure and contains a typical secondary structure. At first, mussel protein is placed 0.5 nm above the surface of the two antifouling membranes (shown in Figure 1), and six chloride ions are randomly placed as counterions to balance the positive charge of the protein. The same thickness of water layer is added above the two polymer membranes. In order to eliminate the possible high energy caused by conformational overlap, the water molecules in the range of 0.2 nm around the protein and polymer membranes are deleted [17,18].

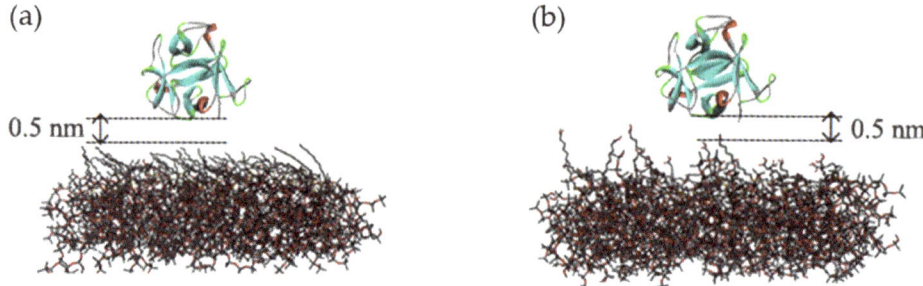

Figure 1. Initial configuration of system: (**a**) protein on the surface of CH3-SAM membrane; (**b**) protein on the surface of COOH-SAM membrane.

In the simulation, the united-atomic force field GROMOS 45a3 is selected [19] and the software package GROMACS (version 4.5.5), is carried out to run the molecular dynamic calculation [20,21]. First of all, for the initial configuration, the steepest descent method was performed several hundred steps to eliminate conformational overlap; then, the NVT ensemble was run for at least 25 ns to obtain the equilibrium of system; then, another 75 ns MD simulation was run to find out the statistical information about mussel protein and the antifouling membrane. During the simulation, in order to reduce the simulation time, the PDMS layers are fixed, and the periodic boundary conditions in XYZ directions are used. For the solvent water, the single point charge (SPC) model is selected [22]. In the simulation, the PME method was used to handle the long-range electrostatic interaction [23], and the radius of non-bond interaction was 1.2 nm. The Berendsen method was used to control the temperature [24], and the LINCS method was selected to constraint the bond of molecule [25]. The simulated step was 2 fs, and the trajectory of system was stored each 100 ps. In the production of simulation, GROMACS analysis program is used to analyze the simulation results, and VMD software (version 1.9.3), is used to visualize the molecular dynamics trajectory.

3. Results and Discussions

3.1. Adsorption Process

In Figure 2, the centroid distance between mussel protein and polymer membrane was calculated, and the variation of minimum distance with time evolution was shown. It is obvious that the distance between mussel protein and polymer membrane for the two systems decreased rapidly and reached equilibrium at a short simulated time, indicating that mussel protein can reach a stable adsorption state for the present simulated model at a short simulated time. We noted that the distance between protein and membrane fluctuated greatly, which indicated that protein constantly adjusted its own configuration during the process of adsorption until an optimal site for adsorption was finally obtained [13–16].

(a) 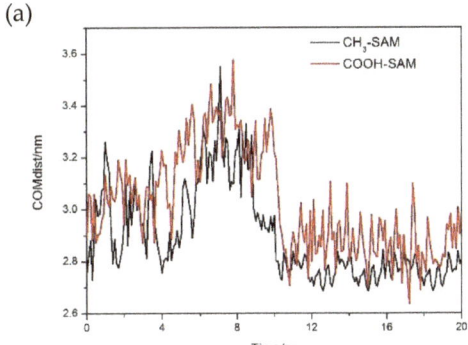 (b)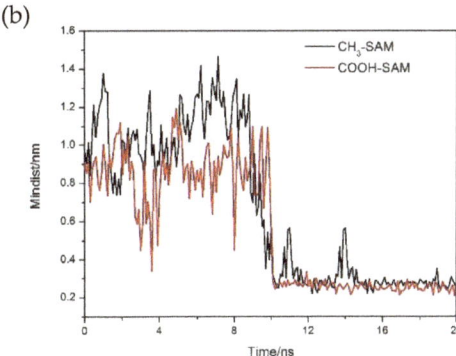

Figure 2. (a) The variation of the distance between mussel protein core and antifouling membrane with time; (b) the variation of the minimum distance between mussel protein and antifouling membrane with time.

After the adsorption equilibrium of protein was obtained, the residue types of amino acid of protein on the CH_3-SAM and COOH-SAM polymer antifouling membrane can be divided, and the results are shown in Table 1. By comparing the residue types of mussel proteins on different self-assembled membrane surfaces, we noted that the nonpolar residues are major on the surface of CH_3-SAM membrane, while the polar residues are major on the surface of COOH-SAM membrane (Figure 3). We speculate that the surface of protein contains hydrophilic polar residues, and the hydrophobic nonpolar residues are mostly in its interior of spherical protein. When the protein interacts with the CH_3-SAM membrane, the hydrophobic residues can be turned over from the interior and form the better combination between the hydrophobic CH_3-SAM surface and hydrophobic residues of protein. In order to prove the speculation, we calculated the interaction energy between mussel protein and two self-assembly membranes, respectively. As shown in Figure 4, it is obvious that the non-bond interaction energy between CH_3-SAM membrane and mussel protein is greater than that between COOH-SAM membrane and the protein. This shows that the hydrophobic surface has a stronger effect on mussel protein molecules, and the adsorption of mussel protein on its surface is more stable and difficult to be separated. At the same time, we noted that the proportion of van der Waals (VDW) interaction is much larger than that of Coulomb interaction during the simulation process. For CH_3-SAM surface, the VDW energy is about 320 kJ/mol, which is approximately 94% contribution to the total energy. While the VDW energy for COOH-SAM surface is about 180 kJ/mol, contributing to 90% to the total energy. These indicate that the driving force of protein adsorption is mainly van der Waals interaction between protein and membrane [26].

Table 1. Statistics of residue types of mussel protein on polymer membrane surface.

System	Polar (%)	Nonpolar (%)
CH$_3$-SAM	21.05 ± 0.05	78.95 ± 0.05
COOH-SAM	71.42 ± 0.05	28.75 ± 0.05

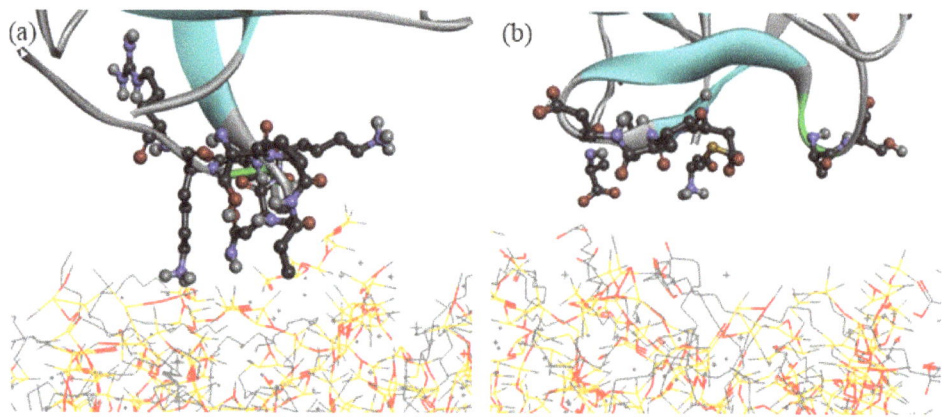

Figure 3. Partial enlarged drawing after stable adsorption. (**a**) CH$_3$-SAM surface of self-assembled membrane. (**b**) COOH-SAM membrane surface.

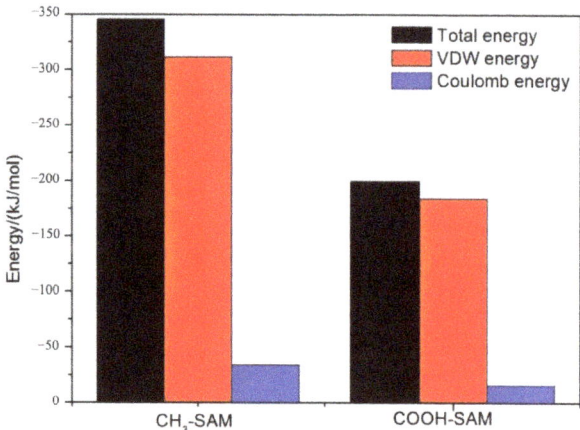

Figure 4. Energy diagram of non-bond interaction between mussel protein and substrate.

3.2. Properties of Hydration Layer on Membrane Surface

The antifouling ability of materials is closely related to the surface hydration layer [4,27]. The hydration layer can act as a physical barrier when the protein is close to the surface of material, and we evaluate the antifouling ability of the two-polymer membrane by analyzing the structure and stability of the hydration layer.

Firstly, the water molecules in the range of 0.4 nm on the surface of membrane were defined as the hydration layer. The mean square displacements (MSD) of water molecules in hydration layer with time evolution for the two investigated systems are shown in

Figure 5. By fitting the two curves and calculating their slopes, the diffusion coefficient (D) of water molecules in the hydration layer can be calculated by Equation (1):

$$D = \frac{1}{2dN} \lim_{t \to \infty} \frac{d}{dt} \sum_{i=1}^{N} \langle [\vec{r_i}(t) - \vec{r_i}(0)]^2 \rangle \quad (1)$$

where N represents the number of target molecules in the system, $\vec{r_i}(0)$ and $\vec{r_i}(t)$ represent the coordinates of the ith particle at 0 and t, respectively.

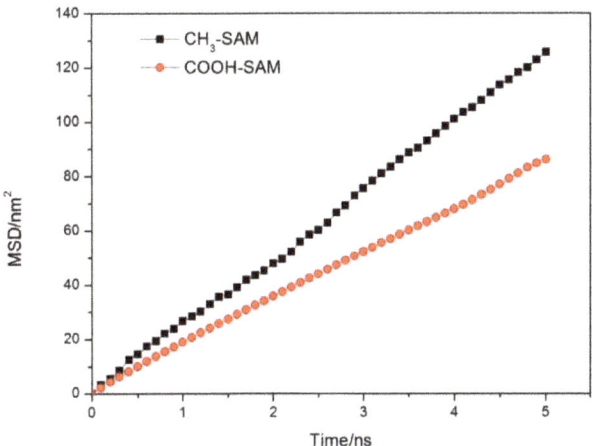

Figure 5. Relation diagram of the variation of average azimuthal shift with time of self-assembled membrane surface hydration layer molecules.

Table 2 lists the Ds of water molecules in the hydration layer and bulk phase. It was found that the D of water molecules in the hydration layer decreased compared with that in the bulk solution for the two systems. It indicates that the interaction between the surface of materials and water molecules restricted the diffusion of water molecules on the membrane surfaces, especially for the COOH-SAM system. It also shows that the binding ability of COOH-SAM self-assembly membrane to water molecules is relatively stronger. Meanwhile, we also noted that the D of water molecules in the vertical direction is much lower than that in the horizontal direction, indicating that the solvent layer molecules are difficult to separate from the surface of the self-assembled membrane.

Table 2. Some property parameters of solvent layer molecules on the substrate surface.

System	Diffusion Coefficients (Ds) × 10^{-5} (cm² s⁻¹)			HBs Life (ps)	HBs Num (nm²)	τ_μ (ps)
	D	$D_\perp$	$D_{//}$			
CH$_3$-SAM	3.21 ± 0.20	0.24 ± 0.43	3.99 ± 0.01	74.34	0.13	46.78
COOH-SAM	2.56 ± 0.09	0.19 ± 0.43	3.73 ± 0.13	129.88	0.25	68.82
Bulk water	3.66 ± 0.04	3.58 ± 0.19	3.69 ± 0.04	-	-	23.93

Relaxation time can describe the limiting ability of antifouling membrane to the molecules of hydration layer. The longer the relaxation time is, the stronger the binding ability of antifouling membrane to water molecules is and representing the better antifouling effect. Its value can be obtained by fitting autocorrelation function [28–30]:

$$C_r(t) = \frac{1}{N_w} \sum_{j=1}^{N_w} \frac{\langle P_{Rj}(0) P_{Rj}(t) \rangle}{\langle P_{Rj}(0) \rangle^2} \quad (2)$$

where P_{Rj} represents a binary operator, if the target molecule j remains in the initial range at time t, then $P_{Rj(t)} = 1$, if not, then $P_{Rj(t)} = 0$. N_w is the total number of target molecules in the initial range defined by us, < > represents ensemble average.

Figure 6 shows the relationship of $C(t)$ and the time t. It can be seen that the autocorrelation functions of water molecules on the two membrane surfaces show the same trend of decay, and the decay of water molecules on CH_3-SAM membrane surface is relatively fast. Fitting the curve using the equation $C_r(t) = A_r \exp(-t/\tau_\mu)$, the relaxation time τ_μ can be calculated. Table 2 lists the τ_μ of hydrated layer molecules and bulk water for the different self-assembled membranes. Due to the existence of polymer membrane, the relaxation time of the surface hydration layer molecules is longer than that of the bulk phase water. It explains that both polymer membranes have limiting effects on the surface hydration layer molecules. The COOH-SAM membrane has greater limiting effect on the hydration layer molecules. Meanwhile, the number and life of hydrogen bonds (HBs) formed between water molecules and self-assembled membrane are also listed in Table 2. The data showed that the number of HBs formed between water molecules and COOH-SAM polymer is relatively stronger, and the life of HBs is relatively long. These also explained the reason why water molecules and COOH-SAM polymer can form the strong hydrogen bonding structure.

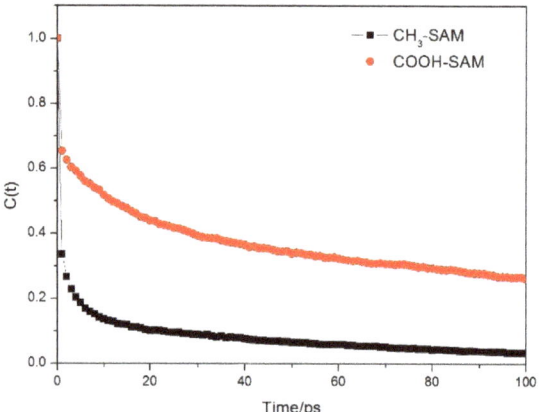

Figure 6. Autocorrelation function of molecules in hydration layer on polymer membrane surface.

3.3. Adsorption Mechanism

In the aqueous environment, a close hydration layer can be formed between mussel protein and antifouling membrane. When the protein molecules in the aqueous approach to the membrane surface, they must destroy the hydration layer first, that is, the adsorption of protein molecules on the antifouling membrane surface is essentially the competitive adsorption behavior between protein molecules and water molecules on the interface. As shown in Figure 7, during the adsorption process, the mussel protein first exposes the hydrophobic residues to the surface through its own structure changes. In this process, the exposure of hydrophobic residues damaged the hydration layer on the surface of the protein. When the protein touches the hydration layer on the membrane surface, in order to complete the adsorption, the energy barrier brought by the hydration layer of the antifouling membrane must be overcome. Due to the different hydrophilicity of antifouling membrane surface, the structure and properties of the surface hydration layer are different. For the hydrophilic carboxyl self-assembled membrane, because the surface contains hydrophilic functional groups, the interaction with water molecules is stronger, and the formed hydration layer is also tighter, so the energy barrier that they should overcome in the process of mussel protein adsorption is larger, which is not conducive to the combination of protein molecules and membrane. However, for the hydrophobic methyl self-assembled membrane, the interaction between the surface and water molecules is weaker, and the formed hydration layer is relatively loose. The mussel protein can be adsorbed on the surface of the antifouling membrane by overcoming the smaller energy barrier, and they form a more stable combination through the hydrophobic interaction.

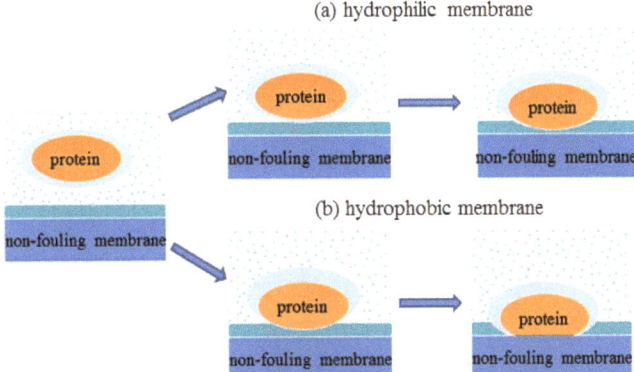

Figure 7. Antifouling mechanism. (**a**) Adsorption process of protein on hydrophilic membrane surface. (**b**) Adsorption process of protein on hydrophobic membrane surface.

4. Conclusions

The adsorption behavior of mussel protein on the surfaces of two antifouling materials was studied by molecular dynamics simulation. By analyzing the adsorption process, including the distance between the protein centroid and the membrane, the type of residues near the adsorption site, the interaction energy between the protein and the antifouling membrane, the diffusion properties of the hydration layer molecules on the membrane surface and the life of HBs, the following simulated conclusions are listed:

(1) In the process of protein adsorption on the surface of different materials, influenced by the chemical composition and structure of the material surface, it will deform through the rotation of its own skeleton, so as to separate the hydration layer on the surface from the protein and form a stable binding with the material surface at the optimal site.

(2) The interaction between mussel protein and antifouling membrane is mainly van der Waals interaction, and the binding between mussel protein and methyl self-assembled membrane is relatively stable.

(3) When mussel protein is adsorbed on the surface of carboxyl self-assembled membrane, it needs to overcome the energy barrier brought by the dense hydrated layer polarized on the surface of the membrane. Compared with the methyl self-assembled membrane, it has better antifouling performance.

In conclusion, this paper uses molecular dynamics method to compare and study the adsorption process of mussel protein on the surface of COOH-SAM membrane and CH_3-SAM membrane and reveals the factors that hydrophilic self-assembled antifouling membrane has better antifouling characteristics from the molecular level, which is of great significance for optimizing and designing new antifouling coatings.

Funding: This research received no external funding.

Institutional Review Board Statement: Not applicable.

Informed Consent Statement: Not applicable.

Data Availability Statement: The data presented in this study is available upon reasonable request.

Conflicts of Interest: The author declares no conflict of interest.

References

1. Clare, A.S.; Evans, L.V. Marine Biofouling: Introduction. *Biofouling* **2000**, *16*, 81–82. [CrossRef]
2. Puleo, D.A.; Rena, B. *Biological Interactions on Materials Surfaces: Understanding and Controlling Protein, Cell, and Tissue Response*; Springer: New York, NY, USA, 2009; pp. 1–17.
3. Dee, K.C.; Puleo, D.A.; Bizios, R. *An Introduction to Tissue-Biomaterial Interactions*; John Wiley & Sons: Hoboken, NJ, USA, 2002.
4. Zheng, J.; Li, L.; Tsao, H.-K.; Sheng, Y.-J.; Chen, S.F.; Jiang, S.Y. Strong repulsive forces between protein and oligo (ethylene glycol) self-assembled monolayers: A molecular simulation study. *Biophys. J.* **2005**, *89*, 158–166. [CrossRef] [PubMed]
5. Sowmi, U.; Michael, P. Analysis of Cooperativity and Group Additivity in the Hydration of 1,2-Dimethoxyethane. *J. Phys. Chem. B* **2021**, *125*, 1660–1666.
6. Vanderah, D.J.; La, H.; Naff, J.; Silin, V.; Rubinson, K.A. Control of protein adsorption: Molecular level structural and spatial variables. *J. Am. Chem. Soc.* **2004**, *126*, 13639–13641. [CrossRef]
7. Liu, Y.L.; Zhang, Y.X.; Ren, B.P.; Sun, Y.; He, Y.; Cheng, F.; Xu, J.X.; Zheng, J. Molecular dynamics simulation of the effect of carbon space lengths on the antifouling properties of hydroxyalkyl acrylamides. *Langmuir* **2019**, *35*, 3576–3584. [CrossRef] [PubMed]
8. Cedervall, T.; Lynch, I.; Foy, M.; Berggard, T.; Donnelly, S.C.; Cagney, G.; Linse, S.; Dawson, K.A. Detailed identification of plasma proteins adsorbed on copolymer nanoparticles. *Angew. Chem. Int. Ed.* **2007**, *46*, 5754–5756. [CrossRef]
9. Aggarwal, P.; Hall, J.B.; McLeland, C.B.; Dobrovolskaia, M.A.; McNeil, S.E. Nanoparticle interaction with plasma proteins as it relates to particle biodistribution, biocompatibility and therapeutic efficacy. *Adv. Drug Deliv. Rev.* **2009**, *61*, 428–437. [CrossRef]
10. Choi, W.; Jin, J.; Park, S.; Kim, J.-Y.; Lee, M.-J.; Sun, H.; Kwon, J.-S.; Lee, H.; Choi, S.-H.; Hong, J. Quantitative interpretation of hydration dynamics enabled the fabrication of a zwitterionic antifouling surface. *ACS Appl. Mater. Interfaces* **2020**, *12*, 7951–7965. [CrossRef]
11. Zheng, H.-R.; Wang, X.-W.; Lin, X.-H.; Geng, Q.; Chen, X.; Dai, W.-X.; Wang, X.-X. Promoted Effect of Polyethylene Glycol on the Photo-Induced Hydrophilicity of TiO_2 Films. *Chim. Sin.-Acta Phys.* **2012**, *28*, 1764–1770.
12. Lüsse, S.; Arnold, K. The interaction of poly (ethylene glycol) with water studied by 1H and 2H NMR relaxation time measurements. *Macromolecules* **1996**, *29*, 4251–4257. [CrossRef]
13. Peng, C.; Liu, J.; Zhao, D.; Zhou, J. Adsorption of hydrophobin on different self-assembled monolayers: The role of the hydrophobic dipole and the electric dipole. *Langmuir* **2014**, *30*, 11401–11411. [CrossRef]
14. Liu, J.; Liao, C.; Zhou, J. Multiscale simulations of protein G B1 adsorbed on charged self-assembled monolayers. *Langmuir* **2013**, *29*, 11366–11374. [CrossRef]
15. Yu, G.; Liu, J.; Zhou, J. Mesoscopic coarse-grained simulations of lysozyme adsorption. *J. Phys. Chem. B* **2014**, *118*, 4451–4460. [CrossRef] [PubMed]
16. Zhao, D.; Peng, C.; Zhou, J. Lipase adsorption on different nanomaterials: A multi-scale simulation study. *Phys. Chem. Chem. Phys.* **2015**, *17*, 840–850. [CrossRef] [PubMed]
17. Caballero-Herrera, A.; Nordstrand, K.; Berndt, K.D.; Nilsson, L. Effect of urea on peptide conformation in water: Molecular dynamics and experimental characterization. *Biophys. J.* **2005**, *89*, 842–857. [CrossRef] [PubMed]
18. Yu, X.; Wang, Q.; Lin, Y.; Zhao, J.; Zhao, C.; Zheng, J. Structure, orientation, and surface interaction of Alzheimer amyloid-β peptides on the graphite. *Langmuir* **2012**, *28*, 6595–6605. [CrossRef]
19. Schuler, L.D.; Daura, X.; Van Gunsteren, W.F. An improved GROMOS96 force field for aliphatic hydrocarbons in the condensed phase. *J. Comput. Chem.* **2001**, *22*, 1205–1218. [CrossRef]
20. Van Der Spoel, D.; Lindahl, E.; Hess, B.; Groenhof, G.; Mark, A.E.; Berendsen, H.C. GROMACS: Fast, flexible, and free. *J. Comput. Chem.* **2005**, *26*, 1701–1718. [CrossRef]
21. Hess, B.; Kutzner, C.; Van Der Spoel, D.; Lindahl, E. GROMACS 4: Algorithms for highly efficient, load-balanced, and scalable molecular simulation. *J. Chem. Theory Comput.* **2008**, *4*, 435–447. [CrossRef] [PubMed]
22. Berendsen, H.J.C.; Postma, J.P.M.; van Gunsteren, W.F.; Hermans, J. *Interaction Models for Water in Relation to Protein Hydration*; Pullman, B., Ed.; Springer: Dordrecht, The Netherlands, 1981; pp. 331–342.
23. Essmann, U.; Perera, L.; Berkowitz, M.L. A smooth particle mesh Ewald method. *J. Chem. Phys.* **1995**, *103*, 8577–8593. [CrossRef]
24. Berendsen, H.J.C.; Postma, J.P.M.; van Gunsteren, W.F.; DiNola, A.; Haak, J.R. Molecular dynamics with coupling to an external bath. *J. Chem. Phys.* **1984**, *81*, 3684–3690. [CrossRef]
25. Hess, B.; Bekker, H.; Berendsen, H.J.C.; Fraaije, J.E.M. LINCS: A linear constraint solver for molecular simulations. *J. Comput. Chem.* **1997**, *18*, 1463–1472. [CrossRef]
26. Dismer, F.; Hubbuch, J. A novel approach to characterize the binding orientation of lysozyme on ion-exchange resins. *J. Chromatogr. A* **2007**, *1149*, 312–320. [CrossRef]
27. Clifton, L.A.; Paracini, N.; Hughes, A.V.; Lakey, J.H.; Steinke, N.-J.; Cooper, J.K.; Gavutis, M.; Skoda, M.W.A. Self-Assembled fluid phase floating membranes with tunable water interlayers. *Langmuir* **2019**, *35*, 13735–13744. [CrossRef]
28. Shao, Q.; He, Y.; White, A.D.; Jiang, S.Y. Difference in hydration between carboxybetaine and sulfobetaine. *J. Phys. Chem. B* **2010**, *114*, 16625–16631. [CrossRef] [PubMed]
29. Liu, Y.L.; Zhang, D.; Ren, B.P.; Gong, X.; Liu, A.; Chang, Y.; He, Y.; Zheng, J. Computational investigation of antifouling property of polyacrylamide brushes. *Langmuir* **2020**, *36*, 2757–2766. [CrossRef] [PubMed]
30. Hayashi, T.; Tanaka, Y.; Koide, Y.; Tanaka, M.; Hara, M. Mechanism underlying bioinertness of self-assembled monolayers of oligo(ethyleneglycol)-terminated alkanethiols on gold: Protein adsorption, platelet adhesion, and surface forces. *Phys. Chem. Chem. Phys.* **2012**, *14*, 10196–10206. [CrossRef] [PubMed]

MDPI
St. Alban-Anlage 66
4052 Basel
Switzerland
Tel. +41 61 683 77 34
Fax +41 61 302 89 18
www.mdpi.com

Molecules Editorial Office
E-mail: molecules@mdpi.com
www.mdpi.com/journal/molecules

www.ingramcontent.com/pod-product-compliance
Lightning Source LLC
LaVergne TN
LVHW070600100526
838202LV00012B/525